49 181 $47.40

9006

FUNDAMENTAL INTERACTIONS IN PHYSICS

Studies in the Natural Sciences

A Series from the Center for Theoretical Studies
University of Miami, Coral Gables, Florida

Volume 1 — IMPACT OF BASIC RESEARCH ON TECHNOLOGY
    Edited by Behram Kursunoglu and Arnold Perlmutter • 1973

Volume 2 — FUNDAMENTAL INTERACTIONS IN PHYSICS
    Edited by Behram Kursunoglu, Arnold Perlmutter,
    Steven M. Brown, Mou-Shan Chen, T. Patrick Coleman,
    Werner Eissner, Joseph Hubbard, Chun-Chian Lu,
    Stephan L. Mintz, and Mario Rasetti • 1973

FUNDAMENTAL INTERACTIONS IN PHYSICS

Conference Chairman
Behram Kursunoglu

Conference Secretary and Senior Editor
Arnold Perlmutter

Scientific Secretaries and Editors

Steven M. Brown
Mou-Shan Chen
T. Patrick Coleman
Werner Eissner

Joseph Hubbard
Chun-Chian Lu
Stephan L. Mintz
Mario Rasetti

Center for Theoretical Studies
University of Miami
Coral Gables, Florida

PLENUM PRESS • NEW YORK-LONDON • 1973

Proceedings of the Coral Gables Conference on Fundamental Interactions
January 22-26, 1973

Library of Congress Catalog Card Number 73-84002
ISBN: 0-306-36902-8

© 1973 Plenum Press, New York
A Division of Plenum Publishing Corporation
227 West 17th Street, New York, N.Y. 10011

United Kingdom edition published by Plenum Press, London
A Division of Plenum Publishing Company, Ltd.
Davis House (4th Floor), 8 Scrubs Lane, Harlesden, London, NW10 6SE, England

All rights reserved

No part of this publication may be reproduced in any form without
written permission from the publisher

Printed in the United States of America

PREFACE

The Center for Theoretical Studies of the University of Miami has been the host of annual winter conferences whose content has expanded from the particular topic of symmetry principles in high energy physics to encompass the bases and relationships of many branches of knowledge. The scope of the Tenth Coral Gables Conference on Fundamental Interactions included astrophysics, atomic and molecular physics, fundamental theories of gravitation, of electromagnetism, and of hadrons, gauge theories of weak and electromagnetic interactions, high energy physics, liquid helium physics, and theoretical biology. The range of topics is partially represented by the scientific talks which form this book. The tangible fruits of the conference are these papers; the intangible ones are the changes of outlook which the participants experienced and the new appreciation they gained of the basic unity of all knowledge.

Historically, the early Coral Gables Conferences witnessed the introduction of the concept of the quark and the attempts to formulate a unification of the internal and space-time symmetries of the elementary particles, while later ones were the initial forums for new unified theories of interactions and for the ideas of scaling, light-cone dominance, and partons. As the recent conferences have treated topics in which it was felt that significant progress might occur and useful concepts be proposed as a result of the attention received in these meetings, it is fitting and proper that this, the last of the Conferences on Fundamental Interactions, is generating a new, more general series of "Coral Gables Orbs Scientiae" to continue this fruitful process in a more capable framework.

The present conference was partially supported by the United States Atomic Energy Commission.

We express our sincere thanks to Mrs. Helga Billings, Mrs. Valerie Casuso, Mrs. Sandra Rice, Miss Rhonda Saathoff, and Mrs. Jacquelyn Zagursky for their kind assistance during the conference and for their excellent work typing these Proceedings.

                              The Editors

TABLE OF CONTENTS

Section One
PARTICLE PHYSICS

Long Range Forces and Broken Symmetries
   P. A. M. Dirac.................................... 1

Fundamental Interactions
   Behram Kursunoglu................................. 19

On SL(6,C) Gauge Invariance
   Abdus Salam....................................... 55

Geometry and Physics of the Elementary Particles
   D. I. Blokhintsev................................. 83

Initial Experiments at NAL
   R. R. Wilson...................................... 97

Total Absorption Detectors and
   Their Applications to High Energy Physics
   Robert Hofstadter................................. 109

Opening Comments on Gauge Theories
   Benjamin Lee...................................... 125

Unified Gauge Theories of Strong,
   Weak and Electromagnetic Interactions
   M. B. Halpern..................................... 129

Concerning the Formulation of Charge
   Independence of the Pion-Nucleon Interaction
   G. Breit.. 143

Theory of a One-Dimensional Relativistic
   Elastic Continuum and Hadronic Wave Equation
   Takehiko Takabayasi............................... 169

An Approach to Hadron Physics Through
  Asymptotic SU(3) and Chiral SU(3) × SU(3)
  Algebra
  S. Oneda and Seisaku Matsuda.................... 175

The Electron and Muon as Eigensolutions
  In Quantum Electromechanics
  Fritz Bopp and W. Lutzenberger................. 183

Section Two
ATOMIC AND MOLECULAR PHYSICS

Slow Positron Collision in Gases
  H. S. W. Massey...................................189

Variational Principles, and Atomic Scattering
  Larry Spruch......................................213

Panel Discussion: Spin and Polarization 229
  Effects in Atomic Processes

Electron-Molecule Scattering
  A. Herzenberg.....................................261

Panel Discussion: Electron-Molecule Collisions ...285

Section Three
VARIOUS TOPICS

X-ray Astronomy
  Herbert Friedman..................................315

Helium in Small Assemblies and Heavy Nuclei
  M. Rasetti and T. Regge...........................337

CONTENTS

Quantum Structural Stability and Biology
  M. Dresden..............................355

List of Participants..........................389

Index...395

P. A. M. Dirac delivered the first paper at the First Decade of the Coral Gables Conferences on Fundamental Interactions.

LONG RANGE FORCES AND BROKEN SYMMETRIES

        P. A. M. Dirac

        Department of Physics

        Florida State University, Tallahassee, Florida

                  and

        Center for Theoretical Studies

        University of Miami, Coral Gables, Florida

    The long range forces in physics are those which fall off according to the inverse square law. The short range forces, on the other hand, fall off exponentially, so that after a very short distance they become effectively zero. The short range forces play an important role in the physics of nuclei and of elementary particles. There are just two long range forces known, the gravitational force and the electromagnetic force. These are the ones that I shall be speaking about today.

    We have a very good theory of the gravitational force which was given to us by Einstein. It explains gravitation in terms of the curvature of space. Now as soon as Einstein brought out this theory people wondered whether the other long range force could not also be explained as some property of space, so that the two forces could be unified. Well, that provided a problem. A solution was very soon put forward by Weyl. I would like to explain Weyl's theory to you, because it is the foundation of what I shall talk about today.

First of all, how are we to understand the curvature of space, as is required for Einstein's theory? Well now if you have a curved two-dimensional space immersed in three dimensions, that's very easy to visualize. You can go on from that to consider a curved space of any number of dimensions immersed in a higher number of dimensions, but it is undesirable to have to introduce a higher number of dimensions, so we consider how we can discuss curvature of a space without increasing the number of dimensions.

One way of doing it is by the notion of parallel displacement of a vector. Suppose we have a vector located at some point in space and we shift it a bit, keeping it parallel to itself as best we can. Suppose we continue shifting it and go on round a loop and come back to our original point. Then if the space is curved we find that our vector is no longer pointing in the original direction. You see that at once if you refer to vectors on a two dimensional curved surface, lying in a three dimensional space, for example, the two dimensional surface of a sphere. You know then that if you take a vector around on the surface of a sphere, moving it parallel to itself, you don't end up with the same direction you started with. If you go around a triangle you get a vector rotated through an angle which is just the angle by which the three angles of a triangle exceed two right angles. That is how one can understand the curvature of space.

Weyl made a generalization of this idea. He said let us consider also the length of a vector and imagine when we shift the vector around a closed loop the final length differs from the original length. Of course, if you just have a curved space immersed in a space of a higher number of dimensions, parallel displacement will give no change in the length of a vector. But Weyl said let us go over to a more general kind of geometry, where the length of a vector changes as well as its direction.

Now how can we set up a mathematical theory where we have these nonintegrable lengths? There's no absolute standard for comparing lengths at two different points. We may take a path joining one point to the other and may shift a vector along that path to the other position and then compare the lengths. But if you shift it along different paths you get different results with that method of comparison. The only way one can talk about lengths of vectors is to set up artificially a standard of length at each point of space and then refer the length to this artificial standard. Then one can discuss the change of length referred to this artificial standard.

Let us take a vector that originally has the length ℓ and is situated at the point with coordinates x^μ. There are four coordinates here, $\mu = 0, 1, 2, 3$. Then we shift this vector and its length changes. It changes by an amount which is proportional to ℓ and is also proportional to the amount of the shift dx^μ. So

$$\delta\ell = \ell \, \kappa_\mu \, dx^\mu \, , \qquad (1)$$

dx_μ is the shift that we are making. The κ_μ are some coefficients which appear in this way. We have these new coefficients κ_μ appearing in our theory, which are to be treated along with the Einstein $g_{\mu\nu}$ which fix the metric.

Now these κ_μ of course are not absolutely defined things, because they depend on our artificial choice of the metric gauge. We might change the metric gauge at each point in space, change it by a factor λ, let us say (λ is a function of the x's). Then we can calculate how κ_μ changes. We find that it changes according to the law

$$\kappa_\mu \rightarrow \kappa_\mu + 1/\lambda \frac{d\lambda}{dx^\mu} = \kappa_\mu + \frac{d\phi}{dx^\mu} \qquad (2)$$

where $\phi = \log \lambda$. Thus the κ_μ's are not quantities having any absolute meaning in our theory. They are subject to these transformations.

If we go around a closed loop and compare the length ℓ of the vector after going around the loop with its original value, the change is $\ell F_{\mu\nu} dS^{\mu\nu}$, where $F_{\mu\nu} = (\frac{\partial \phi_\mu}{\partial x_\nu} - \frac{\partial \phi_\nu}{\partial x_\mu})$ and $dS^{\mu\nu}$ is the area of the loop. Now $\delta\ell/\ell$ is absolute and does not depend on our artificial choice of the metric gauge.

You see this situation is just what we want for describing the electromagnetic field. These κ's we can say are the electromagnetic potentials. They are quantities which don't have an absolute meaning. They are subject to the transformations (2). The quantity $F_{\mu\nu}$ on the other hand is something with an absolute meaning, and gives us the electromagnetic field quantities. The idea of making the lengths of vectors nonintegrable introduces just what we want for describing the electromagnetic field. Now we have a more general geometry than that of Einstein, in which both the gravitational field and the electromagnetic field appear together and are unified.

That was a very beautiful theory, but it was not acceptable to physicists, because it is not in agreement with quantum theory. Up to the present I have said nothing at all about quantum effects, but when you take them into account you realize that there is an absolute standard of length at each point provided by atomic clocks. With atomic clocks we have in the first place a standard for measuring time, and if we take the velocity of light equal to one we get also an absolute standard for measuring distances. Now we might refer the lengths of all our vectors to these standards, and then of course there is no uncertainty in the length of a vector when we shift it. The $\delta\ell$ in (1) is zero when the ℓ at each point in space is referred to atomic clocks. So there is really no place for Weyl's idea when we take into account these quantum effects. The result was that physicists were led to abandon Weyl's theory in spite of its great beauty, and Weyl himself also abandoned it. At most one could consider the changes in length (1) as mathematical in some way, but they don't correspond to reality.

With Weyl's geometry getting abandoned, people retained some of the terminology which was introduced by it. When one changes the metric gauge with the factor λ one calls it a gauge transformation. According to Weyl, gauge transformations are connected with transformations (2) of the potentials. When people abandoned Weyl's theory they retained this terminology, calling (2) a gauge transformation, but the gauge is now something referring only to the electromagnetic potentials and not referring in any way to the metric of space and measurement of lengths of vectors.

With the abandonment of Weyl's theory there remained no real unification of gravitation and the electromagnetic field. One just had to accept that the electromagnetic field is something immersed in the Riemannian space of the Einstein theory and build up one's ideas of physics on that basis. Even so, many physicists were fascinated by the idea of unifying gravitation and the electromagnetic field. It provided one of the major problems of physics which people have been puzzling over for decades. Various solutions have been proposed, but they are all rather artificial and complicated. None of them has the elegance and beauty of Weyl's original theory.

That has been essentially the situation with regard to the long range forces up to the present time except for one further development. This further development refers to some quite different arguments and it leads one to think that the gravitational constant, when measured in atomic units, is not really constant at all but is varying with the age of the universe.

THE LARGE NUMBERS HYPOTHESIS

I'll just outline the argument. If we consider the electric force between the proton and the electron in hydrogen atoms, that is just e^2/r^2. If we consider the gravitational force between the proton and the electron, that would be $Gm_e m_p/r^2$ where m_e is the mass of the electron, m_p is the mass of the proton and G is the gravitational constant. If you take the ratio, it is independent of the distance and is $e^2/Gm_e m_p$. It is a dimensionless number. If we work it out we find that it is somewhere around 10^{39}, an enormous number. Now physicists tend to believe that there should be an explanation for all the dimensionless numbers that turn up, and there should therefore be some explanation for this 10^{39}. How can one ever hope to explain such a very big number?

There is another way of getting a very big number and that is to consider the age of the universe. According to the ideas of the expanding universe, the universe started very small or as a point with some big bang, and the pieces are continuing to separate. Their separation is observed and from their speed one can calculate how long ago the big bang occurred. According to recent estimates it's about 2×10^{10} years, and if we express it in terms of some atomic unit based on the charge and mass of the electron, etc. we get a number around 10^{39}. It's again an extremely large number and about the same as the previous one.

When one sees this connection one feels that it is not just a coincidence; there must be some reason behind it. Some reason which we shall know when we have a better understanding of cosmology and also of atomic theory. From this feeling that these two numbers are connected one is led to suspect that, as time goes on and the second number increases, the first one will also increase in the same ratio. This means that the gravitational constant G expressed in terms of atomic units is varying with the time. Gravitation is getting weaker as time goes on.

Now if you bring this factor into the theory of gravitation you see that the original Einstein theory cannot be maintained. The Einstein theory demands that the gravitational constant shall be a constant. In fact it is equal to one if you use a natural system of units. It certainly cannot have such a variation as this argument requires.

We have to modify the Einstein theory in some way, and a natural way of modifying it is to suppose that the element of distances ds in the Einstein theory, given by $ds^2 = g_{\mu\nu} dx^\mu dx^\nu$, is not the same as the element of distance measured by atomic clocks. Let us call the element of distance in the Einstein theory ds_E and let us suppose that the element of distance measured by atomic clocks is something different, ds_A. The ratio of these two is something which depends on the epoch, the time measured from the origin of creation. With these two ds's we shall have the gravitational constant varying with time. I believe we have to make some modification in the Einstein theory on these lines.

Now if we look back on Weyl's theory we see that the objection to it vanishes. Weyl's ideas are to be applied to ds_E. Atomic clocks measure ds_A which is something different. Now ds_A remains invariant when we take it round a closed loop, but it could very well be that ds_E does change when we take it around a closed loop, and that it changes in accordance with the geometry of Weyl.

Well, that is the theory that I want to propose to you today. We should reintroduce Weyl's theory. It is such a beautiful theory and it provides such a neat way of unifying the long range forces. And there is really no clash with atomic ideas when we have the two ds's. That sets one looking up the old work on Weyl's theory, which is more than 50 years old.

DEVELOPMENT OF WEYL'S THEORY

We have to insert into Einstein's theory these transformations of the metric gauge and see how various quantities vary under these gauge transformations. Let us take first the $g_{\mu\nu}$ which come into the definition of distance. If we make the transformation of gauge in which our lengths are multiplied by λ, then $g_{\mu\nu}$ gets multiplied by λ^2, because the dx's don't change when we change the metric gauge and ds has to get multiplied by λ. We have here a quantity which is multiplied by λ^2 when we make this transformation of the gauge.

We shall be dealing continually with quantities which get multiplied by some power of λ when we change the gauge. If we take a general quantity X, let us say, which goes over into $X\lambda^n$ when we make the change in the gauge, we say that X is of power n. That's the definition of the power of any quantity. If we have a scalar or a tensor, it may change according to this law. We then say that it is a co-scalar or co-tensor of power n. If n is zero we call it an in-scalar or an in-tensor. An in-scalar is something which is unaffected by transformations of coordinates and also by transformations of gauge. There is complete invariance with in-scalars.

In the Einstein theory, if we have any tensor T and we differentiate it with respect to x^μ, the result is not a tensor. But we can modify the derivative and get something, which is called the covariant derivative, which is a tensor. This modified derivative will be denoted by $T_{:\mu}$.

Now if the original tensor was a co-tensor and we form its covariant derivative, the result is not in general a co-tensor again. But we can make a further modification in it, bringing in the potentials, and get something which is co-tensor, which we call the co-covariant derivative, and denote by $T_{*\mu}$. We thus have a process of differentiation which is applicable with Weyl's geometry, and can be applied to any co-tensor to give another co-tensor.

Now we can proceed to develop the theory along parallel lines with the development of Einstein's theory. I shall not go into the details, but I'll just say that in the Einstein theory we have a curvature tensor with four suffixes $B_{\mu\nu\rho\sigma}$ which tells us how vectors change when they are carried around a closed loop. There's a corresponding thing in Weyl's geometry, which we may denote by $*B_{\mu\nu\rho\sigma}$. Now one gets Einstein's law of gravitation by contracting two of the suffixes in B and putting the result equal to zero. The contraction gives us $B_{\mu\nu\rho}{}^{\rho} = R_{\mu\nu}$, say which one puts equal to zero for empty space to get Einstein's law of gravitation.

We can do the corresponding thing with Weyl's geometry, putting $*B_{\mu\nu\rho}{}^{\rho} = *R_{\mu\nu}$. Then you might think that the law of empty space in Weyl's geometry would be just $*R_{\mu\nu} = 0$. Everything would then be analogous with the new geometry. But is would not be satisfactory to take $*R_{\mu\nu} = 0$ as our law of gravitation, because we would like to have our field equations derivable from an action principle. If we can set up an action principle, then we can take into account interaction with other things.

THE ACTION PRINCIPLE

Now how do we get an action principle? Ordinarily one takes the action I equal to the integral of some scalar quantity, let us call it Ω:

$$I = \int \Omega \sqrt{(-g)}\, d^4 x \quad ,$$

where g is the determinant of all the $g_{\mu\nu}$'s. Then this quantity I is an invariant in the Einstein theory, and if we choose a suitable Ω we get Einstein's law of gravitation. In fact we have to just take Ω equal to $R_\mu{}^\mu = R$ say, the total curvature.

Now what would be the corresponding process with Weyl's geometry? In the Weyl geometry we need the action to be an in-invariant. It has to be invariant not only under changes of coordinates but also under changes of gauge. Now $\sqrt{(-g)}$ is of power 4, since each $g_{\mu\nu}$ is of power of 2. If you multiply four of them together to get the determinant and then take the square root, the result is of power 4. Ω we need to be of power -4 in order to have an in-invariant I. So to get a satisfactory scheme of equations in Weyl's geometry, we must find a co-scalar of power -4 and use that to provide us with an action integral.

Now let us look at the electromagnetic field. The usual expression for the action of the electromagnetic field is $1/4\, F_{\mu\nu} F^{\mu\nu}$. The $F_{\mu\nu}$ agree with the $F_{\mu\nu}$ we had before and provide a co-tensor. Now it turns out that this expression is just of power -4. $F_{\mu\nu}$ is of power zero. Then you have to raise two suffixes to get $F^{\mu\nu}$. Raising each suffix brings in a power of -2, because g with two suffixes upstairs is of power -2 to correspond with g with two suffixes downstairs being of power +2. So this electromagnetic action thus has just the right power to be included in our action integral.

Now how about the gravitational part of the action? $*R_{\mu\nu}$ is of power zero and $*R = *R_\mu{}^\mu$ is thus of power -2. If we just put into Ω the term $*R$ to correspond to the R which we have in the Einstein theory, that would not do. It would not give us an in-invariant action. So some further complication is needed. Weyl saw this difficulty, and he proposed that one should take the square of *R in the action, so that one should have

$$\Omega = 1/4 F_{\mu\nu} F^{\mu\nu} + k(*R)^2 \quad ,$$

where k is some numerical coefficient. That is the action that Weyl proposed. But is excessively complicated. This curvature quantity *R itself is quite complicated, and if you square it you get an excessively complicated thing. That was a further reason which led to the abandonment of Weyl's theory.

You must do something else if you want to get a reasonable theory. What you might do is to take *R = 0 as a constraint in the theory, and bring gravitation in by this constraint. If you do that, when it does not matter what the power of *R is. To bring in a constraint equation like that, we must add to Ω the term γ *R; so

$$\Omega = 1/4 F_{\mu\nu} F^{\mu\nu} + \gamma * R$$

where γ is a Lagrangian multiplier. We have the Lagrangian multiplier appearing in the theory as another scalar field quantity. It has to be of power -2 in order that γ *R may be of power -4 and give us an in-invariant action integral.

Now we have rather gone over to the scalar-tensor theory of gravitation as proposed by Brans and Dicke (1961). We see the need for introducing a scalar quantity γ to describe the gravitational field along with the usual $g_{\mu\nu}$ tensor quantity, just in order to get an action that is not excessively complicated for use with the Weyl geometry.

So I propose that we should go over to the scalar-tensor theory of Brans and Dicke, but I want to develop it somewhat differently than Brans and Dicke did. In the first place, it is convenient to replace γ by β^2. The quantity β is then a co-scalar of power -1. Now if you have a scalar field quantity appearing in the action, you can very well introduce further quantities involving its derivatives. The further quantities have to be co-scalars of power -4, and there is an obvious one: $\beta_{*\mu}\beta^{*\mu}$, where $*\mu$ means the co-covariant derivative in the way I mentioned to you. This quantity is a co-scalar and turns out to be of power -4, so we can add it on with the arbitrary numerical coefficient k. Thus we get the action density

$$\Omega = 1/4 F_{\mu\nu}F^{\mu\nu} + \beta^2 {}^*R + k\beta^{*\mu}\beta_{*\mu} . \qquad (3)$$

If you examine this quantity (3), you will find that there is one choice of the coefficient k which makes it specially simple, namely k = 6. With this value of k, it turns out that the potentials κ disappear from the expression (3). There are κ's appearing in the co-covariant derivatives, and it so happens that they cancel out those in β^2 *R for k = 6. That means an essentially simpler theory for this one value of k.

That is where I depart from Brans and Dicke. Brans and Dicke worked with a general k. I assume this one value of k giving a specially simple action, which I believe describes what we have for the vacuum, when there is no matter present. If there is matter present we just have to add on the appropriate further terms to the action and calculate the resulting equations of motion.

Now if we have the potentials disappearing from our action with only the $F_{\mu\nu}$ remaining, we can make

transformations (2) of the potentials without affecting the action at all. We now have a theory in which we have three kinds of transformation which are possible:
(i)   We can make any transformation of coordinates.
(ii) We can make any transformation of the metric gauge combined with the transformation (2) of the potentials.
(iii) For the vacuum, we can make a transformation of the metric gauge alone or a transformation (2) of the potentials alone.

Now the important thing there is that the third kind of transformation is not a general one. The first two can be made with any kind of matter interacting with the gravitational and the electromagnetic fields. The third transformation can be made when there is no matter present at all, but it cannot be made in general. In general there will be potentials surviving in the expression for the action, which prevent us from making a transformation of the metric gauge alone or a transformation of the potentials alone. The transformations (iii) which can be made for the vacuum or with certain specially simple kinds of matter interacting in the vacuum, but cannot be made in general. Those are the essential features of this theory.

SYMMETRY BREAKING

I would like to say a little about the effect of this theory on symmetry breaking. That is best discussed in terms of a simple example. Suppose we have a particle moving along a definite world line. Let us take a field point close to the world line. Let us take a vector of length ℓ, located at this field point, and let us shift it in a direction roughly parallel to the world line, that is to say we shift it through an amount dx^o, the variation of the time coordinate, assuming the particle to be approximately at rest. Then we have our length changing by the amount

$$\delta\ell = \ell\kappa_o dx^o \quad . \tag{4}$$

Now let us suppose the particle is charged. Then the potential at the field point will consist mainly of the Coulomb potential arising from this charge, because the field point is supposed to be quite close to the particle. The κ_o thus consists of the Coulomb potential. If you look at this formula (4), you see that if you change the sign of the charge, you change the sign of κ_o, and you therefore change the sign of $\delta\ell$. It may be that we start off with a sign of the charge such that ℓ increases when we go into the future. If we change the sign of the charge we have ℓ decreasing when we go into the future. Now there's no symmetry between a thing increasing and a thing decreasing. If the changes are infinitesimal there is a mathematical symmetry, but when you deal with finite changes, there is no symmetry.

The result is that we have a theory in which there is no symmetry between positive and negative charge. That is a striking feature of the Weyl geometry. It is an essential way in which this theory differs from the theory in which the electromagnetic field is supposed to be merely immersed in the geometry needed for the Einstein gravitational theory.

Now suppose we interchange future and past. That means changing the sign of dx^o. That again results in symmetry breaking. The $\delta\ell$ would be positive in one case and negative in the other case. However, if we change the sign of the charge and also interchange future and past, we get back to the original situation.

Now the high energy physicists have introduced some operations referring to the symmetry processes which are important in high energy physics. There is P, the parity operation connected with changing a left-handed system of coordinates to a right-handed system of coordinates. Then there is C, charge conjugation, which changes the sign of the charge. And there's the operation T, which reverses time. Experimentally it is found that all three symmetries are broken, but the product of the three seems to be conserved: PCT is conserved as far as is known.

The Weyl geometry does not give any breaking of the P symmetry. The breaking of the P symmetry must be ascribed to the short range forces. I can't see any way of accounting for it in terms of long range forces. On the other hand, the long range forces handled in this way do lead to a breaking of C and T, with preservation of CT. There we have something coming out from the theory which we were not expecting in the first place, and which does provide a special point of interest.

This breaking of the C and T symmetries does not occur when you are dealing with an action for which transformations of the type (iii) can be applied. In such a case the transformations of the metric gauge are independent of the transformations of the potentials, and under those conditions the breaking of the C symmetry will not show up. The breaking of the C symmetry and of the T symmetry do show up in the equations of motion only if we have a sufficiently complicated kind

of action for which tranformations (iii) are not possible.
This requires a particle of a rather complicated kind.
The simplest kinds of particles don't show it up.

That is perhaps satisfactory because experimentally
the breaking of C and T symmetries is a very rare
event. It shows itself up only in one particular kind
of particle, the K-meson. For all the others C and T
seem to be preserved. According to the present theory
one would say that the K-meson involves an action of the
more complicated kind, such that transformations of type
(iii) are no longer possible. Those are the ideas which
I wanted to speak to you about.

REFERENCE

Brans, C. and Dicke, R. H. (1961) Phys. Rev. 124, 925.

FUNDAMENTAL INTERACTIONS*

Behram Kursunoglu

Center for Theoretical Studies

University of Miami, Coral Gables, Florida 33124

I. THE PHOTON

The concept of the photon and its interaction with matter contains maximum information on the unity of physics and on the fundamentals of all elementary states of matter. Just as the classical electromagnetic field was the basis of special relativity the concept of the photon laid the foundations of quantum theory. The interaction of photons with matter is the best understood subject in physics. We may expect the same particle to play a major role in building a quantum theory of elementary particles. The current theories have demonstrated the crucial role of the ideas of symmetry and in particular of broken symmetries in the analysis of the basic states of matter. Therefore we must formulate the electromagnetic

---

*The mathematical details of the points of view presented here can be found in B. Kursunoglu, Phys.Rev. D. Vo. 1, No. 4, pp. 1115-1132, 15 Feb. 1970 and Phys.Rev. D., Vol. 2, No. 4, pp. 717-735, 15 August 1970.

theory on the basis of broken symmetries. The implimentation of such a program in a unique way can provide a natural basis to discover the broken internal and space-time symmetries for all other massless as well as massive states of matter in the small.

Let me begin with the established symmetries of the photon. Any state of polarization of a photon can be expressed as a superposition of any two orthogonal states of polarization. If a photon is linearly polarized at 45° to the optic axis of some crystal (or a birefrengent substance) this polarization can be represented as a superposition of the x- and y- polarizations of equal amplitude and in phase. If the crystal plate is of just right thickness to introduce a 90° phase shift between the x- and y- polarization, then the photon will come out circularly polarized (right or left). Such a thickness of the crystal plate is referred to as a quarter-plate because it introduces a quarter cycle phase difference between the x- and y- polarizations. This process of changing a state of linear polarization into a state of circular polarization can be represented, as will be shown, by a unitary transformation applied to the given state of polarization. Therefore if we wish we can change a circular (linear) polarization into a linear (circular) polarization. In fact, if a linearly polarized photon is sent through two quarter wave plates, it will come out linearly polarized again but at right angles to the original direction.

A given state of polarization, say the x-polarization can further be split up as a superposition of two orthogonal states of opposite parities. The electric and magnetic components of the photon are associated with different states of parity and are orthogonal.

FUNDAMENTAL INTERACTIONS

Thus in terms of these four orthogonal states we may, for an electromagnetic wave, write

$$|\Psi_{tr}\rangle = \frac{1}{2}[E_+|-1,-1\rangle + E_-|-1,1\rangle + iH_+|1,-1\rangle + iH_-|1,1\rangle], \quad (1.1)$$

where

$$E_\pm = E_1 \pm i E_2, \quad H_\pm = H_1 \pm iH_2,$$

and where the 6-dimensional orthonormal basis vectors $|\pm 1, \pm 1\rangle$ are given by

$$|-1,-1\rangle = \frac{1}{\sqrt{2}}\begin{bmatrix}|-1\rangle \\ |-1\rangle\end{bmatrix}, \quad |-1,1\rangle = \frac{1}{\sqrt{2}}\begin{bmatrix}|1\rangle \\ |1\rangle\end{bmatrix},$$

$$|1,-1\rangle = \frac{1}{\sqrt{2}}\begin{bmatrix}|-1\rangle \\ -|-1\rangle\end{bmatrix}, \quad |1,1\rangle = \frac{1}{\sqrt{2}}\begin{bmatrix}|1\rangle \\ -|1\rangle\end{bmatrix}. \quad (1.2)$$

The vectors $|-1\rangle$, $|1\rangle$ belong to the eigenvalues -1 and $+1$ of the photon spin matrix

$$K_3 = \begin{bmatrix} 0 & -i & 0 \\ i & 0 & 0 \\ 0 & 0 & 0 \end{bmatrix},$$

where

$$|-1\rangle = \frac{1}{\sqrt{2}}\begin{bmatrix}1 \\ -i \\ 0\end{bmatrix}, \quad |1\rangle = \frac{1}{\sqrt{2}}\begin{bmatrix}1 \\ i \\ 0\end{bmatrix},$$

and the spin operator of the photon is given by

$$S_3 = \begin{bmatrix} K_3 & 0 \\ 0 & K_3 \end{bmatrix}. \quad (1.3)$$

The vectors $|\pm 1, -1\rangle$ and $|\pm 1, 1\rangle$ belong to the eigenvalues ± 1 of the parity matrix

$$-\Gamma_{44} = \begin{bmatrix} 0 & -I_3 \\ -I_3 & 0 \end{bmatrix}, \quad I_3 = \begin{bmatrix} 1 & 0 & 0 \\ 0 & 1 & 0 \\ 0 & 0 & 1 \end{bmatrix}. \quad (1.4)$$

The complex 6-dimensional vector (1.1) represents the transverse part of the field. The longitudinal part can be written as

$$|\Psi_{Long.}\rangle = \tfrac{1}{\sqrt{2}} E_3 |-1,0\rangle + \tfrac{1}{\sqrt{2}} i H_3 |1,0\rangle, \quad (1.5)$$

where

$$|-1,0\rangle = \tfrac{1}{\sqrt{2}} \begin{bmatrix} |0\rangle \\ |0\rangle \end{bmatrix}, \quad |1,0\rangle = \tfrac{1}{\sqrt{2}} \begin{bmatrix} |0\rangle \\ -|0\rangle \end{bmatrix}, \quad (1.6)$$

and the vector

$$|0\rangle = \begin{bmatrix} 0 \\ 0 \\ 1 \end{bmatrix}$$

belongs to zero eigenvalues of K_3. The vectors $|\pm 1, 0\rangle$ belong to the eigenvalue 0 of S_3 and the eigenvalues ± 1 of $-\Gamma_{44}$.

The expansion (1.1), in terms of the polarization states alone, using the properties $E_1 = -H_2$, $E_2 = H_1$ of the transverse waves, can be written as

$$|\Psi_{Tr.}\rangle = (E_1 + iH_1)\Gamma_+ |-1,-1\rangle + (E_1 - iH_1)\Gamma_- |-1,1\rangle$$
$$= (E_1 + iH_1)\Gamma_+ |1,-1\rangle - (E_1 - iH_1)\Gamma_- |1,1\rangle, \quad (1.7)$$

FUNDAMENTAL INTERACTIONS

where the projection operators $\Gamma_\pm$ for polarization are given by

$$\Gamma_\pm = \tfrac{1}{2}(1 \pm \Gamma_5), \quad \Gamma_5 = \begin{bmatrix} I_3 & 0 \\ 0 & -I_3 \end{bmatrix}. \qquad (1.8)$$

Hence the right and left circularly polarized states $\Gamma_+|\pm 1,-1\rangle$, $\Gamma_-|\pm 1,1\rangle$ are eigenstates of Γ_5 belonging to the eigenvalues +1 and -1, respectively. Thus, because of the conservation of parity by the electromagnetic interaction the two parity states of the same polarization can be fused into a single state and we obtain the photon as a superposition of two states of polarizations.

In terms of the 6×6 matrix Γ_{44}, as defined in (1.4), the energy density of the electromagnetic field is given by

$$\tfrac{1}{2}(E^2 + H^2) = \tfrac{1}{4} \langle \bar{\psi} | \Gamma_{44} | \psi \rangle, \qquad (1.9)$$

where the adjoint function $\langle \bar{\psi} |$ is defined as

$$\langle \bar{\psi} | = \langle \psi | \Gamma_{44} = [\chi, \chi^*], \quad \chi = \underline{E} + i\underline{H}.$$

The energy momentum tensor

$$T_{\mu\nu} = \tfrac{1}{4} g_{\mu\nu} f^{\rho\sigma} f_{\rho\sigma} - f_{\mu\rho} f_\nu{}^\rho, \qquad (1.10)$$

in terms of the photon state $|\psi\rangle$ can be written as

$$T_{\mu\nu} = \tfrac{1}{4} \langle \bar{\psi} | \Gamma_{\mu\nu} | \psi \rangle \qquad (1.11)$$

where the ten matrices $\Gamma_{\mu\nu}$, with $\Gamma_{44} = \Gamma_{11} + \Gamma_{22} + \Gamma_{33}$, are the generators of the 6-dimensional SU(3,1) transformations which leave an interval of the form

$$|z_4|^2 - |z_1|^2 - |z_2|^2 - |z_3|^2 \quad ,$$

invariant. We may also define an energy momentum density 4-vector or a photon current density by

$$P_\mu = \frac{1}{4c} <\bar{\psi}|\Gamma_{\mu\nu}v^\nu|\psi> \quad , \qquad (1.12)$$

where the unit vector

$$v_\mu = \frac{dx_\mu}{ds} \quad , \quad v^\mu v_\mu = 1 \quad ,$$

is a solution of the Lorentz's equations of motion of an electric charge in an external field $f_{\mu\nu}$, viz.,

$$\frac{dv_\mu}{ds} = - \frac{e}{mc^2} f_\mu{}^\nu v_\nu \quad . \qquad (1.13)$$

Hence it follows that at any given time τ the vector $v_\mu(\tau)$ can be obtained from its previous value at time τ_o by a continuous unfolding in time of a Lorentz transformation induced by the external field $f_{\mu\nu}$ as

$$v(\tau) = \Lambda(\tau,\tau_o)v(\tau_o) \quad , \quad s=c\tau \quad , \qquad (1.14)$$

where the Lorentz matrix $\Lambda(\tau,\tau_o)$ is given by

$$\Lambda(\tau,\tau_o) = \exp\left[- \frac{ie}{2mc^2} \int_{s_o}^{s} (f^{\mu\nu}M_{\mu\nu})ds\right] \quad , \qquad (1.15)$$

and $M_{\mu\nu}$ are the usual six 4×4 matrices generating 4-dimensional Lorentz transformations and where we have the obvious relation

$$- \frac{1}{2}i(f^{\rho\sigma}M_{\rho\sigma})_\mu{}^\nu = f_\mu{}^\nu \quad .$$

FUNDAMENTAL INTERACTIONS

The statement (1.14) also defines a gauge transformation at a point $x_\mu(\tau)$ or at time $\tau(x_\mu)$.

The current density P_μ as defined by (1.12) transforms as a 4-vector under Lorentz transformations of the coordinates given by

$$x'_\mu = \Lambda_\mu^\nu x_\nu ,$$

which requires the transformation of $|\psi\rangle$ according to

$$|\psi\rangle = \exp[-\frac{ie}{2mc^2} \int_{s_0}^{s} (f^{\mu\nu} J_{\mu\nu}) ds] |\psi\rangle = S|\psi\rangle , \quad (1.16)$$

and of the matrices $\Gamma_{\mu\nu}$ according to

$$S^\dagger \Gamma_{\mu\nu} S = \Lambda_\mu^\rho \Lambda_\nu^\sigma \Gamma_{\rho\sigma} . \quad (1.17)$$

The six 6×6 matrices $J_{\mu\nu}(=-J_{\nu\mu})$ together with the nine matrices $\Gamma_{\mu\nu}$ saturate the Lie algebra of the group SU(3,1). The total energy and momentum of the field is given by

$$P_\mu = \int P_\mu \, d\sigma = \int T_{\mu\nu} v^\nu d\sigma , \quad (1.18)$$

where the invariant volume element $d\sigma$ is defined as

$$d\sigma = v_\mu d\sigma^\mu ,$$

and the $d\sigma_\mu$ are the 3-dimensional surface elements bounding a 4-dimensional region of space-time. For $e - 0$ the vector P_μ is, of course, conserved

$$\partial^\mu P_\mu = 0 . \quad (1.19)$$

The function $\Psi[x_\mu(s)]$ and the particle velocity v_μ in the definition of P_μ are computed at time s to obtain $P_\mu[x_\rho(s)]$. The form of the current density P_μ as a function of Ψ_a (a=1,2,...,6) and velocity v_μ implies that we must consider mixed quantities $v_\mu \Psi_a$ for the description of the electromagnetic field which contains particle and field variables. This can be achieved by absorbing the 4-vector v_μ into $|\Psi\rangle$ and $\langle\bar{\Psi}|$ and writing the definition (1.12) of P_μ, uniquely, in the form

$$P_\mu = -\frac{i}{8c} \langle\bar{\phi}|\Gamma_\mu|\phi\rangle , \qquad (1.20)$$

where the four 30×30 matrices Γ_μ are defined by

$$\Gamma_\mu = \Gamma_{\mu\nu} \beta^\nu , \qquad (1.21)$$

and where the new "wave function" $|\phi\rangle$ is given, in index notation by

$$(|\phi\rangle)_{a\lambda} = v_\lambda \psi_a , \quad \lambda = 1,2,\ldots,5,$$

$$\langle\bar{\phi}| = \langle\phi|\beta_o \Gamma_{44} , \quad v_5 = 1 .$$

The four 5×5 matrices β_μ are the usual Duffin-Kemmer matrices and together with the six matrices

$$\beta_{\mu\nu} = i(\beta_\mu \beta_\nu - \beta_\nu \beta_\mu) \qquad (1.22)$$

saturate the Lie algebra of the group SO(3,2). The 5-dimensional matrix β_o is represented by

FUNDAMENTAL INTERACTIONS

$$\beta_o = [g_{\lambda\omega}] = \begin{bmatrix} -1 & 0 & 0 & 0 & 0 \\ 0 & -1 & 0 & 0 & 0 \\ 0 & 0 & -1 & 0 & 0 \\ 0 & 0 & 0 & 1 & 0 \\ 0 & 0 & 0 & 0 & 1 \end{bmatrix} . \quad (1.23)$$

Furthermore the β_μ satisfy the relations

$$\beta_\mu \beta_\nu \beta_\rho + \beta_\rho \beta_\nu \beta_\mu = g_{\mu\nu} \beta_\rho + g_{\nu\rho} \beta_\mu . \quad (1.24)$$

Another important 5-dimensional matrix is β_5 obeying the algebraic relations

$$[\beta_{\mu\nu}, \beta_5] = 0 , \quad [\beta_o, \beta_5] = 0 , \quad \beta_\mu \beta_5 + \beta_5 \beta_\mu = 0 , \quad (1.25)$$

and having the form

$$\beta_5 = \begin{bmatrix} I_4 & 0 \\ 0 & -1 \end{bmatrix} \quad (1.26)$$

where I_4 is 4×4 unit matrix. Furthermore we have

$$\beta_o \beta_\mu \beta_o = \beta_\mu^\dagger . \quad (1.27)$$

The projection operators

$$\beta_\pm = \tfrac{1}{2}(1 \pm \beta_5) \quad (1.28)$$

act on $|\phi\rangle$ to produce

$$|\phi_-\rangle = \beta_-|\phi\rangle = \begin{bmatrix} 0 \\ 0 \\ 0 \\ 0 \\ |\psi\rangle \end{bmatrix}, \quad |\phi_+\rangle = \beta_+|\phi\rangle = \begin{bmatrix} v_1|\psi\rangle \\ v_2|\psi\rangle \\ v_3|\psi\rangle \\ v_4|\psi\rangle \\ 0 \end{bmatrix}. \quad (1.29)$$

From the conservation law (1.19) it follows, by differentiation, that

$$\partial^\mu(\langle\bar\phi|)\Gamma_\mu|\phi\rangle + \langle\bar\phi|\Gamma_\mu\partial^\mu|\phi\rangle = 0 \quad.$$

This equation is valid for an arbitrary $|\phi\rangle$ satisfying the equation

$$\Gamma_\mu\partial^\mu|\phi\rangle = 0, \quad (1.30)$$

or its hermitian conjugate form

$$\partial^\mu\langle\bar\phi|)\Gamma_\mu = 0 \quad. \quad (1.31)$$

The equation (1.30) contains the generators $\Gamma_{\mu\nu}$ of the group SU(3,1) and the generators β_μ of the group SO(3,2) where $\Gamma_{\mu\nu}$ and β_μ do, of course, commute. The equation (1.30), by the use of the projection operators (1.28), can be broken up into the two equations

$$\Gamma_\mu\partial^\mu|\phi_-\rangle = 0, \quad (1.32)$$

and

$$\Gamma_\mu\partial^\mu|\phi_+\rangle = 0 \quad. \quad (1.33)$$

The equation (1.32) reduces to

$$\Gamma_{\mu\nu}\partial^\nu|\psi\rangle = 0, \quad (1.34)$$

FUNDAMENTAL INTERACTIONS

and in 4-dimensional space time notation we obtain

$$\partial_\rho f_{\mu\nu} + \partial_\mu f_{\nu\rho} + \partial_\nu f_{\rho\mu} + g_{\mu\rho} \partial^\sigma f_{\nu\sigma} - g_{\rho\nu} \partial^\sigma f_{\mu\sigma} = 0 \quad, \tag{1.35}$$

which contracted with respect to $g^{\mu\rho}$ or $g^{\nu\rho}$, yields Maxwell's equations

$$\partial_\rho f_{\mu\nu} + \partial_\mu f_{\nu\rho} + \partial_\nu f_{\rho\mu} = 0 \quad, \quad \partial^\nu f_{\mu\nu} = 0 \quad. \tag{1.36}$$

The equation (1.33) by the use of (1.32) or (1.36) reduces to

$$i\hbar \frac{d}{d\tau} |\Psi\rangle = H |\Psi\rangle \quad, \tag{1.37}$$

where

$$H = ic J_{\mu\nu} v^\mu p^\nu, \quad p^\mu = i\hbar \partial^\mu \quad,$$

and

$$\frac{d}{ds} = v^\mu \partial_\mu = \frac{1}{i\hbar} v_\mu p^\mu \quad.$$

The equation (1.37) describes the fact that $|\Psi\rangle$ is a function of the world line of the particle represented by $x_\mu = x_\mu(\tau)$ and does therefore obey a dynamical equation involving the total time derivative of $|\Psi\rangle$. In fact, in the rest frame of the particle moving in the field $|\Psi\rangle$ we have

$$\frac{d}{ds} |\Psi\rangle = \frac{\partial}{\partial x_4} |\Psi\rangle + v \cdot \nabla |\Psi\rangle \to \frac{1}{c} \frac{\partial}{\partial t} |\Psi\rangle$$

and

$$H = ic J_{\mu\nu} v^\mu p^\nu \to -ic J_{4j} p_j = c\Gamma_5 K \cdot p \quad,$$

where

$$J_{4j} = i\,\Gamma_5 K_j = \begin{bmatrix} iK_j & 0 \\ 0 & -iK_j \end{bmatrix}.$$

Hence for $v=0$ the equation (1.37) reduces to the Schrödinger form of Maxwell's equations which is also obtained for $\mu=4$ from (1.34).

The eigenvalues of H which follow from the algebraic equation

$$H(H^2 - E_o^2) = 0$$

are given by $\pm E_o$ and 0 where

$$E_o = c\,v^\mu p_\mu\,, \quad p_\mu p^\mu = 0\,. \qquad (1.38)$$

However because of the Maxwell's equations (1.34) the eigenvalue 0 does not occur since it corresponds to the solutions in the direction of the photon momentum. Therefore the operator H, as long as it acts on a state $|\Psi\rangle$ satisfying (1.34), has an inverse. We can use H and introduce a quantity by

$$|\Psi\rangle = \sqrt{(16\pi H)}\,|\eta\rangle\,, \qquad (1.39)$$

where $|\eta\rangle$, having the dimensions of $(\text{Volume})^{-\frac{1}{2}}$, will be interpreted as the photon wave function. The eigenvalue E_o of H which follows from (1.38) represents the energy of the photon as observed by an observer moving with a velocity v with respect to an emitting particle. The physical meaning of E_o can be seen by writing it as

$$E_o = \hbar\omega_o = \hbar\omega\,\frac{1 \mp v\cdot\hat{p}}{\sqrt{(1-\frac{v^2}{c^2})}}\,, \quad \omega = \frac{\omega_o\sqrt{(1-\frac{v^2}{c^2})}}{1 \pm v\cdot\hat{p}}\,,$$

FUNDAMENTAL INTERACTIONS   31

where $\omega = cp$. Thus the frequency ω_o is emitted by a stationary atom oscillating at one of its natural frequencies which in this case is ω_o. The frequency ω observed by the moving observer is given with a minus sign in the denominator if the observer is moving closer to the emitting atom and plus sign if the observer is moving away from the atom. This is the usual Doppler effect.

The emergence of the wave equation (1.37) is not surprising because the value of the external field at the particle itself i.e. $f_{\mu\nu}[x(s)]$ will vary according to the trajectory of the particle and hence $f_{\mu\nu}$ must change as a function of τ (the proper time of the photon as observed by the particle). Such a change is observed, in the rest frame of the particle, as solutions of the Maxwell's equations (1.34).

The equations (1.34) consist of $4 \times 6 = 24$ equations for six variables E and H. This apparent overdetermination of the field arises from the fact that there are two unitary operations of changing or reshuffling the states of polarization and parity. Thus one has, besides $|\psi\rangle$, the states $\Gamma_5|\psi\rangle$, $P|\psi\rangle$, and $P\Gamma_5|\psi\rangle$. We have, therefore because of the conservation of these symmetries, four kinds of states describing the same photon and hence the reason for the 24 equations in (1.34). However, if, for example, parity were not conserved then the states $|\psi\rangle$ and $P|\psi\rangle$ could not describe the same particle We have thus established the fact that a free photon is a $6 \times 5 = 30$ dimensional representation of the group $G_o = SU(3,1) \times SO(3,2)$ in, of course, a Lorentz covariant way.

Now the equations (1.34) for $\mu=4$ and $\mu=1,2,3$ can be written as

$$(p_4 + \Gamma_5 K \cdot p)\Gamma_{44}|\psi\rangle = 0 , \qquad (1.40)$$

$$[p_j+p_4K_j\Gamma_5+i(p\times K)_j]\Gamma_{44}|\psi> = 0 \quad , \qquad (1.41)$$

where in deriving (1.41) we have used (1.40). The two equations (1.40) and (1.41) yield the result

$$p_\mu p^\mu = p_4^2 - p^2 = 0 \quad , \quad p_4 = \pm |p| \quad . \qquad (1.42)$$

In terms of the unit vector $\hat{p} = \dfrac{1}{p_4}|p|$ we may rewrite (1.40) and (1.41) as

$$|\gamma> = -\Gamma_5 K\cdot\hat{p}|\gamma> \quad , \qquad (1.43)$$

$$\hat{p}|\gamma> = -(\Gamma_5 K + i\hat{p}\times K)|\gamma> \qquad (1.44)$$

where

$$|\gamma> = -\Gamma_{44}|\psi> \quad .$$

By using photon polarization operators Γ_+ and choosing the photon momentum p as the z-direction of the coordinates we may express (1.43) and (1.44) in the forms

$$|\gamma> = -K_3|\gamma_+> + K_3|\gamma_-> \quad , \qquad (1.45)$$

$$K_+|\gamma_+> = 0 \quad , \quad K_-|\gamma_-> = 0 \quad , \qquad (1.46)$$

where

$$K_\pm = K_1 \pm iK_2 \quad , \quad |\gamma_\pm> = \Gamma_\pm|\gamma>$$

and K_1, K_2, K_3 are the photon spin matrices given by

$$K_1 = \begin{bmatrix} 0 & 0 & 0 \\ 0 & 0 & -i \\ 0 & i & 0 \end{bmatrix}, \quad K_2 = \begin{bmatrix} 0 & 0 & i \\ 0 & 0 & 0 \\ -i & 0 & 0 \end{bmatrix}, \quad K_3 = \begin{bmatrix} 0 & -i & 0 \\ i & 0 & 0 \\ 0 & 0 & 0 \end{bmatrix}.$$

$$(1.47)$$

FUNDAMENTAL INTERACTIONS 33

The $|\gamma_\pm\rangle$ describe two orthogonal opposite states of polarizations. In deriving the above relations we have used the definitions

$$\Gamma_{\mu\nu} = \begin{bmatrix} 0 & B^*_{\mu\nu} \\ B_{\mu\nu} & 0 \end{bmatrix}, \qquad (1.48)$$

where

$$B_{44} = B_{11} + B_{22} + B_{33} = I_3 \ , \quad B_{4j} = B_{j4} = K_j \ ,$$

$$B_{j\ell} = K_j K_\ell + K_\ell K_j - I_3 \delta_{j\ell} \ . \qquad (1.49)$$

Hence

$$\Gamma_{4j} = \begin{bmatrix} 0 & -K_j \\ K_j & 0 \end{bmatrix} = \Gamma_{44} \Gamma_5 K_j \ , \qquad (1.50)$$

$$\Gamma_{j\ell} = \Gamma_{44} B_{j\ell} \ ,$$

and

$$\Gamma_5 \Gamma_{\mu\nu} + \Gamma_{\mu\nu} \Gamma_5 = 0 \ . \qquad (1.51)$$

The equations (1.46) yield the relations

$$E_1 = -H_2, \quad E_2 = H_1, \quad E_3 = H_3 = 0 \qquad (1.52)$$

for a monochromatic electromagnetic wave, so that the invariants $E^2 - H^2$ and $E \cdot H$ vanish. The results (1.48), because of (1.46) and (1.45), imply the expansion (1.1) for a photon. The transverse and longitudinal parts of the field can be expressed as

$$|\Psi_{tr.}\rangle = B_+|\psi\rangle \ , \quad |\Psi_{Long.}\rangle = B_-|\psi\rangle(=0) \ , \qquad (1.53)$$

where the projection operators $B_\pm$ are defined by

$$B_\pm = \frac{1}{2}(1 \pm \Gamma_{44}\Gamma_{33}), \quad \Gamma_{44}\Gamma_{33} = \begin{bmatrix} B_{33} & 0 \\ 0 & B_{33} \end{bmatrix}, \quad B_{33} = \begin{bmatrix} 1 & 0 & 0 \\ 0 & 1 & 0 \\ 0 & 0 & -1 \end{bmatrix}. \quad (1.54)$$

We shall see that in the presence of charges the longitudinal parts E_3 and H_3 of the field are coupled to matter along with the transverse components E_1, E_2, H_1, H_2.

The states of circular polarizations can be transformed into the states of linear polarizations by applying a unitary transformation given by

$$|\Psi_{LP}\rangle = U|\psi_{cp}\rangle, \quad (1.55)$$

where the 6×6 unitary matrix U is given by

$$U = \frac{1}{\sqrt{2}} \begin{bmatrix} \langle -1|, & \langle -1| \\ \langle 1|, & \langle 1| \\ \langle 0|, & \langle 0| \\ \langle -1|, & -\langle -1| \\ \langle 1|, & -\langle 1| \\ \langle 0|, & -\langle 0| \end{bmatrix}, \quad (1.56)$$

where we have used the vectors (1.6).

Finally we characterize a photon state of a given momentum by a unit polarization vector $|a\rangle$ expressed as

$$|a\rangle = b_{-1}|e_1\rangle + b_1|e_2\rangle + c_{-1}|e_3\rangle + c_1|e_4\rangle, \quad (1.57)$$

where

$$b_{-1} = \frac{1}{\sqrt{2}} e^{i\alpha} \cos\gamma, \quad c_{-1} = -(-1)^n \frac{1}{\sqrt{2}} e^{-i\alpha} \cos\gamma,$$

$$b_1 = \frac{1}{\sqrt{2}} e^{i\beta} \sin\gamma, \quad c_1 = (-1)^n \frac{1}{\sqrt{2}} e^{-i\beta} \sin\gamma, \quad (1.58)$$

FUNDAMENTAL INTERACTIONS 35

$$|b_{-1}|^2 + |b_1|^2 + |c_{-1}|^2 + |c_1|^2 = 1 \quad , \qquad (1.59)$$

and where n=0 and n=1 correspond to the states of even and odd parities. The quantities $|b_{-1}|^2$, $|b_1|^2$, $|c_{-1}|^2$, $|c_1|^2$ represent probabilities of definite polarization and parity of the photon determined by the unit vectors $|e_\mu\rangle$, $\mu = 1,2,3,4$, given by

$$|e_1\rangle = \begin{bmatrix}1\\0\\0\\0\\0\\0\end{bmatrix}, \quad |e_2\rangle = \begin{bmatrix}0\\1\\0\\0\\0\\0\end{bmatrix}, \quad |e_3\rangle = \begin{bmatrix}0\\0\\1\\0\\0\\0\end{bmatrix}, \quad |e_4\rangle = \begin{bmatrix}0\\0\\0\\0\\1\\0\end{bmatrix}. \qquad (1.60)$$

Because of the normalization condition (1.59) and also the arbitrariness of the phase of $|a\rangle$ we can choose $\alpha = 0$ and thereby describe the state of polarization by two real parameters β and γ for the even and odd parity states of the photon. The expansion (1.57) can now be written as

$$|a\rangle = \tfrac{1}{\sqrt{2}}[\cos\gamma|e_1\rangle + e^{i\beta}\sin\gamma|e_2\rangle - (-1)^n\cos\gamma|e_3\rangle + (-1)^n e^{-i\beta}\sin\gamma|e_4\rangle]. \qquad (1.61)$$

Thus $\beta=0$ and $n=0,1$ denote linear polarizations at an angle γ to the x-axis with even and odd parity, respectively. The values $\gamma = \tfrac{\pi}{4}$, $\beta = \pm\tfrac{\pi}{2}$, $n=0,1$ denote right and left circular polarizations of even and odd parities represented by the vectors $|\pm 1, \pm 1\rangle$. Arbitrary values of β and γ correspond to elliptic polarizations of either parity. Furthermore, from $\Gamma_{4\mu} p^\mu \Gamma_{\pm} |\Psi\rangle = 0$ we get the helicities

$$\frac{K \cdot p}{p_4} \Psi_{\pm} = \pm \Psi_{\pm} \quad ,$$

which correspond to right and left circular polarizations.

Now, the photon equation (1.32) can, by the action of polarization operators $\Gamma_\pm$, be split up into the two equations

$$\Gamma_\mu \partial^\mu \Gamma_+ |\phi_-\rangle = 0 \;,\; \Gamma_\mu \partial^\mu \Gamma_- |\phi_-\rangle = 0 \;, \qquad (1.62)$$

describing right and left circularly polarized photons. Under the parity operation effected by the unitary operator

$$P = -\beta_o \Gamma_{44} I_s \;, \qquad (1.63)$$

the equations for the right and left circularly polarized photons transform into one another. This is due to the conservation of parity and it means that both equations in (1.62) describe one and the same photon in its two different states of polarization.

Now in the case of electromagnetic interaction the equation (1.30) is replaced by

$$\Gamma_\mu \partial^\mu |\phi\rangle = Q_{\mu\nu} \beta^\mu J^\nu |v\rangle \;, \qquad (1.64)$$

where the numerical tensor coefficients $Q_{\mu\nu}(=-Q_{\nu\mu})$ serve to relate the complex and real representations of the electromagnetic field according to

$$f_{\mu\nu} = \frac{1}{2} Q_{\mu\nu a} \Psi_a \;, \qquad (1.65)$$

and where $|v\rangle$ is the column vector form of the velocity v_λ with $v_5=1$ and J_μ is the electric current density. As in (1.32) and (1.33) the equation (1.64) can be split up according to

FUNDAMENTAL INTERACTIONS

$$\Gamma_\mu \partial^\mu |\phi_\pm\rangle = Q_{\mu\nu}\beta^\mu J^\nu |v_\pm\rangle , \qquad (1.66)$$

where

$$|\phi_\pm\rangle = \beta_\pm |\phi\rangle , \quad |v_\pm\rangle = \beta_\pm |v\rangle .$$

In 4-dimensional notation (1.66) reduces to

$$\partial_\rho f_{\mu\nu} + \partial_\mu f_{\nu\rho} + \partial_\nu f_{\rho\mu} = 0 , \quad \partial^\nu f_{\mu\nu} = J_\mu , \qquad (1.67)$$

and

$$i\hbar \frac{d}{d\tau} |\psi\rangle = H|\psi\rangle + i\hbar Q_{\mu\nu} v^\mu J^\nu , \qquad (1.68)$$

respectively. Under the parity, time reversal, and charge conjugation operations

$$P = -\beta_o \Gamma_{44} I_s , \quad T = -\beta_o \beta_5 CI_t , \quad G = \Gamma_{44} \beta_5 C , \qquad (1.69)$$

respectively, the equation (1.30) remains unchanged. However the equation (1.64) gains, under G operation, a minus sign and therefore is not invariant. This is a very interesting result meaning that the equation (1.64) does not contain a correct description of the interaction even though it yields Maxwell's equations (1.67). The use of the complex object $|\phi\rangle$ in (1.64) and the operator (1.69) is not completely equivalent to Maxwell's equations in the presence of a source J_μ. This result is due to the fact that the equation (1.64) contains, in addition to the field $|\psi\rangle$, the particle variable v_μ obeying Lorentz's equations of motion. Therefore, charge conjugation must take into consideration the fact that the same operation is being applied in the equations of motion and therefore the fact that one has two equations of motions for the two signs of the electric charge.

In order to remedy the above situation we must re-examine the symmetry basis of our theory. In fact the answer is contained in the observation that the commutation relations of the group $SU(3,1)$ are satisfied not only by $\Gamma_{\mu\nu}$ but also by $-\Gamma_{\mu\nu}$. Hence in order to restore charge conjugation invariance equation (1.64) must be replaced by

$$q_3 \Gamma_\mu \partial^\mu |\phi\rangle = Q_{\mu\nu} \beta^\mu J^\nu |v\rangle , \qquad (1.70)$$

where

$$q_3 = \begin{bmatrix} 1 & 0 \\ 0 & -1 \end{bmatrix} \qquad (1.71)$$

and does, of course, commute with all the space-time transformations contained in the group G_o as well as in the Lorentz group of four dimensions. The matrix q_3 operates on the two velocity vectors $v_{1\mu}$, $v_{2\mu}$ corresponding to two signs of the electric charge. Thus both $|\psi\rangle$ and $|v\rangle$ have charge indices through the vectors v_μ^ε, where $\varepsilon = 1,2$. We may regard the eigenvalues of q_3 as internal quantum numbers associated with the dynamics of the two signs of electric charges.

We can introduce the charge projection operators

$$q_\pm = \tfrac{1}{2}(1 \pm q_3) \qquad (1.72)$$

and split up the equation (1.70) into the equations

$$\Gamma_\mu \partial^\mu |\phi^+\rangle = Q_{\mu\nu} \beta^\mu J^\nu |v^+\rangle , \qquad (1.73)$$

FUNDAMENTAL INTERACTIONS

$$\Gamma_\mu \partial^\mu |\phi^-\rangle = - Q_{\mu\nu} \beta^\mu J^\nu |v^-\rangle \qquad (1.74)$$

where

$$|\phi^{\pm}_-\rangle = q_{\pm} |\phi\rangle$$

corresponds to the eigenvalues ± 1 of q_3 and where the superscripts $+$ and $-$ correspond to the upper and lower components of $|\phi\rangle$ and $|v\rangle$ in (1.70).

The charge conjugation operation which leaves the electromagnetic interaction invariant is now given by

$$G_I = q_1 \Gamma_{44} \beta_5 C \;, \qquad (1.75)$$

where the matrix q_1 is given by

$$q_1 = \begin{bmatrix} 0 & 1 \\ 1 & 0 \end{bmatrix} \qquad (1.76)$$

and anticommutes with q_3. Hence the 6-dimensional representations $\pm \Gamma_{\mu\nu}$ of the group SU(3,1) are associated with the dynamics of the electromagnetic field and provide us with an internal quantum number represented by q_3. The electromagnetic interactions, because of the two signs of the electric charge, are not invariant under an SU(2) transformation of the general form $\exp(\frac{1}{2} i q \cdot \alpha)$, where the components of the vector q are the matrices q_1, q_2, q_3. However, the interaction is invariant under a rotation around the q_2-axis (the neutral axis). We may, at this point, state the fact that the electromagnetic interactions are born whenever the photon interaction with the charges breaks the symmetry operations contained in the finite dimensional representations of the group

$$G_I^\circ = SU(3,1) \times SO(3,2) \times SU(2) \;, \qquad (1.77)$$

where only 6-, 5- and 2- dimensional representations of SU(3,1), SO(3,2) and SU(2) are included, respectively.

We can now take an additional step and combine the cases of $e \neq 0$ and $e = 0$ into an internal space and write in place of the equations (1.30) and (1.70) the equation

$$\Gamma_\mu \partial^\mu |\phi\rangle = T_3 \, Q_{\mu\nu} \beta^\mu J^\nu |v\rangle \quad , \tag{1.78}$$

where

$$T_3 = \begin{bmatrix} 1 & 0 & 0 \\ 0 & -1 & 0 \\ 0 & 0 & 0 \end{bmatrix} \quad , \tag{1.79}$$

and the corresponding charge conjugation operator is now given by

$$G_I = T_1 \Gamma_{44} \beta_5 C \quad , \tag{1.80}$$

where

$$T_1 = \begin{bmatrix} 0 & 1 & 0 \\ 1 & 0 & 0 \\ 0 & 0 & 0 \end{bmatrix} . \tag{1.81}$$

The matrices T_1, T_3 and

$$T_2 = \begin{bmatrix} 0 & -i & 0 \\ i & 0 & 0 \\ 0 & 0 & 0 \end{bmatrix} \quad , \tag{1.82}$$

generate 3-dimensional transformations of the SU(2) sub-group of the group SU(3).

A further generalization of the above theory follows from introducing a one-dimensional gauge group U(1) into the SU(3,1) since the commutation relations for SU(3,1) are also satisfied by the matrices of the form

$$T_3 \Gamma_{\mu\nu} + Y\rho g_{\mu\nu} = \Lambda_{\mu\nu} , \qquad (1.83)$$

where ρ is a non-negative real parameter and the operator Y commutes with all the space-time transformations and also with T_3, viz.,

$$[T_3, Y] = 0 . \qquad (1.84)$$

The hermitian diagonal matrix Y generates gauge transformation of the first kind

$$|\psi> = \exp(i\rho Y)|\psi> . \qquad (1.85)$$

The matrices $\Lambda_{\mu\nu}$ as defined by (1.84), contain all the possible algebraic solutions of the commutation relations for the group SU(3,1). The operator Y can assume positive and negative integer values and also the value Y=0.

The one-dimensional representations of U(3,1) emerge whenever T_3 assumes zero eigenvalue. The SU(3,1) representations, which include the regular 4-dimensional representation, belong to Y = 0 eigenvalue in $\Lambda_{\mu\nu}$.

II. WEAK AND STRONG INTERACTIONS

Massive particles with spin $\frac{1}{2}$ and spin 0 and 1 are described by Dirac and Kemmer wave equations containing the generators of the group SO(3,2) corresponding to its 4-dimensional representations for Dirac and 5- and 10-dimensional representations for the Kemmer equation. A massless particle like, for example, the photon does not fit into the Kemmer scheme since a zero mass Kemmer equation for spin 1 does not yield Maxwell's equations. The same situation, for entirely different reasons than in the case of Kemmer equation, prevails for the Dirac equation since setting mass equal to zero neither yields a two-component neutrino nor predicts the existence of two neutrinos.

According to the theory discussed above the wave equation for the photon by involving the generators of SU(3,1) and SO(3,2) breaks a more detailed symmetry than the Dirac and Kemmer equations. Instead of setting the mass equal to zero in the Dirac and Kemmer equations we may generalize the photon wave equation (1.30) or rather its decomposed forms (1.32) and (1.33), by introducing a mass term in them. Thus an equation describing massive states is of the form

$$(\Lambda_\mu p^\mu - mc\gamma\rho)|\psi\rangle = 0 \qquad (2.1)$$

where the 4 matrices Λ_μ are given by

$$\Lambda_\mu = T_3 \Gamma_{\mu\nu} \beta^\nu + Y\rho\beta_\mu \quad ,$$

with the β's assuming both 5- and 10-dimensional representations of SO(3,2), for mesons and

$$\Lambda_\mu = T_\mu \Gamma_{\mu\nu} \gamma^\nu + Y\rho\gamma_\mu \qquad (2.2)$$

for baryons. For Y or $\rho = 0$ we obtain, for the 6-dimensional representation of SU(3,1), the wave equation of the photon. The equation (2.1) is invariant with respect to P, G and T operations.

FUNDAMENTAL INTERACTIONS 43

Now all the representations of U(3,1) can be generated from its 4-dimensional regular representation. Using the symbol ☐ for the vector index μ and ☐ (vertical) for the fully antisymmetric indices [μνρ], the Young Tableaux yield

$$4 \times 4 = \square \times \square = \square\square \;+\; \begin{array}{c}\square\\\square\end{array} = 10 + 6 ,$$

$$\bar{4} \times 4 = \begin{array}{c}\square\\\square\\\square\end{array} \times \square = \begin{array}{c}\square\\\square\\\square\\\square\end{array} + \begin{array}{c}\square\square\\\square\\\square\end{array} = 1 + 15 .$$

Based on the association of the photon symmetries with the 6-dimensional representation of SU(3,1) we shall assume that the basic massive states correspond to the representations

$$6 = \begin{array}{c}\square\\\square\end{array}$$

$$6 \times 6 = \begin{array}{c}\square\\\square\end{array} \times \begin{array}{c}\square\\\square\end{array} = \begin{array}{c}\square\\\square\\\square\\\square\end{array} + \begin{array}{c}\square\square\\\square\\\square\end{array} + \begin{array}{c}\square\square\\\square\square\end{array} = 1 + 15 + 20 ,$$

$$6 \times 6 \times 6 = 3\begin{array}{c}\square\\\square\end{array} + \square\square + \begin{array}{c}\square\square\\\square\square\end{array} + \begin{array}{c}\square\square\square\\\square\square\square\end{array} + 2\begin{array}{c}\square\square\\\square\square\\\square\end{array} =$$

$$6 + 6 + 6 + 10 + \overline{10} + 50 + 64 + 64 , \text{ etc.}$$

(2.3)

where $\overline{10}$ is the conjugate representation to the N=10 representation of SU(3,1).

The extension of (1.32) and (1.33) to massive systems of integral spin can be written as

$$T_3 \Gamma_{\mu\nu} \beta^\mu p^\nu |\phi_{\underline{+}}\rangle + \gamma\rho(\beta^\mu p_\mu - mc)|\phi\rangle = 0 , \qquad (2.4)$$

where

$$|\phi_{\underline{+}}\rangle = \beta_{\underline{+}} |\phi\rangle .$$

For half-integral spin we write

$$-iT_3 \Gamma_{\mu\nu} \gamma^\mu p^\nu |\psi_\pm\rangle + Y\rho(\gamma^\mu p_\mu - mc)|\psi\rangle = 0 , \qquad (2.5)$$

where

$$|\psi_\pm\rangle = \gamma_\pm |\psi\rangle ,$$

$$\gamma_\pm = \tfrac{1}{2}(1\pm\gamma_5), \quad \gamma_5 = \gamma_1\gamma_2\gamma_3\gamma_4 = \begin{bmatrix} 0 & I_2 \\ I_2 & 0 \end{bmatrix} ,$$

$$\gamma_4 = \beta , \quad \gamma_j = \begin{bmatrix} 0 & \sigma_j \\ -\sigma_j & 0 \end{bmatrix} , \quad \alpha_j = \begin{bmatrix} 0 & \sigma_j \\ \sigma_j & 0 \end{bmatrix} ,$$

$$\gamma_\mu\gamma_\nu + \gamma_\nu\gamma_\mu = 2 g_{\mu\nu} ,$$

$$\gamma_4^\dagger = \gamma_4 , \quad \gamma_j^\dagger = -\gamma_j .$$

$$\sigma_{\mu\nu} = \tfrac{1}{2} i(\gamma_\mu\gamma_\nu - \gamma_\nu\gamma_\mu) .$$

$$\sigma_{4j} = i\alpha_j, \quad \sigma_{jk} = \varepsilon_{jk\ell}\sigma_\ell , \qquad (2.6)$$

$$(\gamma^\mu p_\mu)^2 = p_\mu p^\mu .$$

The equation (2.1) may be used to obtain the mass spectrum of strongly interacting hadrons which have no weak interactions. For baryons and leptons with weak interactions the appropriate equation, with different values of the ρ, Y pair corresponding to baryons and leptons, is given by (2.5). The mass difference of n and p as well as that of μ and e may be described as being due to the coupling of the parity non-conserving and parity conserving parts,

FUNDAMENTAL INTERACTIONS

the coupling constant being $Y\rho$.

The value $\rho = \infty$ in (2.5) leads to

$$(\gamma^\mu p_\mu - mc)|\psi\rangle = 0, \qquad (2.7)$$

and corresponds to switching off the mass difference of the weakly interacting particles. Thus $|\psi\rangle$ is the wave function for particles of various spins and equal mass.

Setting $\rho=0$ or $Y=0$ in the equation (2.5) yields for the case of the γ_- projection the equation

$$\Gamma_{\mu\nu}\gamma^\mu p^\nu \gamma_- |\nu\rangle = 0. \qquad (2.8)$$

This equation can be split into two equations

$$\Gamma_{\mu\nu}\gamma^\mu p^\nu \Gamma_+ \gamma_- |\nu\rangle = 0, \qquad (2.9)$$

and

$$\Gamma_{\mu\nu}\gamma^\mu p^\nu \Gamma_- \gamma_- |\nu\rangle = 0. \qquad (2.10)$$

Under a parity transformation effected by the operator $P = -\beta\Gamma_{44}I_s$ the equations (2.9) and (2.10) transform into

$$\Gamma_{\mu\nu}\gamma^\mu p^\nu \Gamma_- \gamma_+ |\nu\rangle = 0, \qquad (2.11)$$

and

$$\Gamma_{\mu\nu}\gamma^\mu p^\nu \Gamma_I \gamma_I |\nu\rangle = 0, \qquad (2.12)$$

and hence the equations (2.9) and (2.10) do not conserve parity. This is contrary to the case discussed for the photon where a parity transformation, because of the scalar character of β_5, led to an interchange of the two possible states of the same particle.

Thus the equations (2.9) and (2.10) describe different massless particles. By using the reductions

$$\Gamma_{\mu\nu}\gamma^\mu p^\nu = \tau_o \gamma^\mu p_\mu [(2-3A)\Lambda_+ + 3A\Lambda_-] \quad,$$

$$\Gamma_{\mu\nu}\gamma^\mu p^\nu \gamma_- = \tau_o \gamma^\mu p_\mu [(2-3A)\Gamma_- + 3A\Gamma_+]\gamma_- \quad,$$

$$\Lambda_+ \Gamma_- = \Gamma_-\gamma_- \quad, \quad \Lambda_-\Gamma_- = \Gamma_-\gamma_+ \quad, \quad \Lambda_\pm = \tfrac{1}{2}(1\pm\Gamma_5\gamma_5)$$

of the operator $\Gamma_{\mu\nu}\gamma^\mu p^\nu$ we obtain the equations (2.9) and (2.10) in the form

$$\gamma^\mu p_\mu \Gamma_+ \gamma_- A |\nu\rangle = 0 \quad, \tag{2.13}$$

$$\gamma^\mu p_\mu \Gamma_- \gamma_- (2-3A) |\nu\rangle = 0 \quad, \tag{2.14}$$

where

$$\tau_o = \frac{\Gamma_{\mu\nu} p^\mu p^\nu}{p^2} \quad, \quad \tau_o^2 = 1$$

has been factored out and where

$$A = \tfrac{1}{3}[\tfrac{15}{4} - s(s+1)], \quad s = \tfrac{1}{2}, \quad s = \tfrac{3}{2} \quad.$$

The spin projection operator A ($=A^2$) assumes the value 1 for $s = \tfrac{1}{2}$ and the value 0 for $s = \tfrac{3}{2}$. Therefore by operating on (2.14) by A and $1-A$ we can split it up into two equations,

$$\gamma^\mu p_\mu \Gamma_- \gamma_- A |\nu\rangle = 0 \quad, \tag{2.15}$$

$$\gamma^\mu p_\mu \Gamma_- \gamma_- (1-A) |\nu\rangle = 0 \quad. \tag{2.16}$$

FUNDAMENTAL INTERACTIONS

Now from

$$\gamma^\mu p_\mu \gamma_- = \beta(p_4 + \sigma\cdot p)\gamma_- \ , \quad p_4 = \pm|p| \ , \qquad (2.17)$$

it follows that the equations (2.13), (2.15) and (2.16) can be written as

$$(p_4 + \sigma\cdot p)|\nu_e, \tfrac{1}{2}\rangle = 0 \ , \qquad (2.18)$$

$$(p_4 + \sigma\cdot p)|\nu_\mu, \tfrac{1}{2}\rangle = 0 \ , \qquad (2.19)$$

$$(p_4 + \sigma\cdot p)|\nu_o, \tfrac{3}{2}\rangle = 0 \ , \qquad (2.20)$$

where

$$|\nu_e, \tfrac{1}{2}\rangle = \gamma_-\Gamma_+ A|\nu\rangle \ , \quad |\nu_\mu, \tfrac{1}{2}\rangle = \gamma_-\Gamma_- A|\nu\rangle \ , \qquad (2.21)$$

and

$$|\nu_o, \tfrac{3}{2}\rangle = \gamma_-\Gamma_-(1-A)|\nu\rangle \ . \qquad (2.22)$$

For $s = \tfrac{1}{2}$ the projection operator A strikes out 16 components (the two $s = \tfrac{3}{2}$ states) of the 24-component wave function $|\nu\rangle$ and the projection operator γ_- halfs the remaining 8 components. We now have four components left and they are further divided into two 2-component systems by the action of Γ_- or Γ_+. For $s = \tfrac{3}{2}$ the spin projection operator A strikes out 8 states in $|\nu\rangle$ and the remaining 16 states for $s = \tfrac{3}{2}$ particles are halved by the action of Γ_- and these 8 states are further reduced by half by the action of γ_-. We are thus left with 4 components to describe a spin $\tfrac{3}{2}$ massless particle. We shall see that the wave function (2.22) for $s = \tfrac{3}{2}$ state vanishes and therefore no massless particle of

$s = \frac{3}{2}$ is predicted.

The wave functions (2.21) can be combined into a 4-component wave function

$$|\nu_e, \tfrac{1}{2}\rangle + |\nu_\mu, \tfrac{1}{2}\rangle = \gamma_- A |\nu\rangle = |\nu, \tfrac{1}{2}\rangle$$

which has two upper and two lower components describing two different neutrinos. The electron and muon neutrino wave functions belong to the eigenvalues +1 and -1 of Γ_5. The two neutrinos belong to eigenvalue -1 of γ_5. The positive energy solutions correspond to particles with their spins and momenta anti-parallel and the negative energy solutions to anti-particles with their spins and momenta parallel. They describe left-handed (negative helicity) particles and right-handed (positive helicity) anti-particles. In the case of photon we can interpret the +1 eigenvalue of Γ_5 as the left-handed photon state and the -1 eigenvalue as the right-handed "antiphoton" state. For the neutrinos the +1 eigenvalue of Γ_5 represents left-handed ν_e state and the -1 eigenvalue represents left-handed $\bar{\nu}_\mu$ state. Thus the distinction between ν_e and ν_μ is in their helicities alone: this result is not surprising since ν_μ is represented in the 4-component wave function $|\nu, \tfrac{1}{2}\rangle$ by the two lower components. Thus the ν_e and ν_μ fields are the positive and negative chiral parts of one 4-component field $|\nu, \tfrac{1}{2}\rangle$.

In any weak process the emitted lepton is always polarized along its direction of flight (longitudinal polarization). The degree of longitudinal polarization normally is equal to $\pm \frac{v}{c}$ for the lepton. The form of interaction is such that the lepton tends to be left-handed (spin and momentum antiparallel) while the antilepton tends to be right handed. In particular for the neutrinos $\frac{v}{c} = 1$, the degree of longitudinal polarization (helicity) is always complete.

The weakly interacting massive leptons as well as baryons are described by

$$-i \Gamma_{\mu\nu}\gamma^{\mu}p^{\nu}\gamma_{-}|\Psi\rangle + \rho(\gamma^{\mu}p_{\mu}-mc)|\Psi\rangle = 0 \quad , \quad (2.23)$$

where

$$\rho_{\ell} \ll \rho_{B} \quad , \quad (2.24)$$

and T_3 and Y for $N=6$ representation of $SU(3,1)$ are not needed, since the wave functions corresponding to eigenvalues $+1$ and -1 of these operators are related by the unitary transformation

$$\Psi' = \Gamma_5 \Psi \quad .$$

The $\rho = \infty$ case corresponds to turning off their mass differences and hence their weak behavior. For $\rho=0$ we have $|\psi\rangle = |\nu\rangle$. For $\rho \neq 0$ the equation (2.23) describes particles related by weak decay processes. Therefore the mass difference is due to the coupling of the parity conserving and parity nonconserving parts of the system. The conserved hermitian current density for the equation (2.23) is given by

$$J_{\mu} = \langle\bar{\psi}|(-i\Gamma_{\mu\nu}\gamma^{\nu}\gamma_{-} + \rho\gamma_{\mu})|\psi\rangle \quad , \quad (2.25)$$

where

$$\langle\bar{\psi}| = \langle\psi|\Gamma_{44}\beta \quad . \quad (2.26)$$

The wave equation (2.23) yields the mass formula

$$\frac{m}{M} = \pm \sqrt{(1 \pm \frac{\sqrt{3}}{\rho})} \qquad (2.27)$$

which for $\rho_\ell = 1.73213$ results in a muon electron mass ratio of

$$\frac{M_\mu}{M_e} = \frac{\sqrt{(1+\frac{\sqrt{3}}{\rho_\ell})}}{\sqrt{(1-\frac{\sqrt{3}}{\rho_\ell})}} = 206.77 \quad .$$

For the neutron-proton mass difference we have

$$\rho_B = 1258.66 \quad .$$

III. WEAK COUPLING OF LEPTON AND HADRONS

The set of weak currents are of the form

$$J_\mu^{b\ell o} = \sqrt{G} \, \langle \bar{\ell} | (-i\Gamma_{\mu\nu}\gamma^\nu \gamma_- + \rho_\ell \gamma_\mu) | \nu \rangle \quad ,$$

$$J_\mu^{b\ell o} = \sqrt{G} \, \langle \bar{b} | (-i\Gamma_{\mu\nu}\gamma^\nu \gamma_- + \rho_B \gamma_\mu) | \nu \rangle$$

$$J_\mu^{b\ell} = \sqrt{G} \, \langle \bar{b} | (-i\Gamma_{\mu\nu}\gamma^\nu \gamma_- + \rho \gamma_\mu) | \ell \rangle \quad .$$

By operating with the spin projection operators A and 1-A the wave equation (2.23) can be decomposed into

$$[\rho(\gamma^\mu p_\mu - mc) + i\tau_o \gamma^\mu p_\mu \gamma_-]|\psi_-,\tfrac{1}{2}\rangle + [\rho(\gamma^\mu p_\mu - mc) - 3i\tau_o \gamma^\mu p_\mu \gamma_-]|\psi_+,\tfrac{1}{2}\rangle = 0$$
(2.28)

$$[\rho(\gamma^\mu p_\mu - mc) - 2i\tau_o \gamma^\mu p_\mu \gamma_-]|\psi_-,\tfrac{3}{2}\rangle + \rho(\gamma^\mu p_\mu - mc)|\psi_+,\tfrac{3}{2}\rangle = 0 \quad ,$$
(2.29)

FUNDAMENTAL INTERACTIONS

where

$$|\psi_{\pm},\tfrac{1}{2}\rangle = \Gamma_{\pm}A|\psi\rangle \;,\;\; |\psi_{\pm},\tfrac{3}{2}\rangle = \Gamma_{\pm}(1-A)|\psi\rangle \;. \qquad (2.30)$$

From (2.29) by operating with Γ_{-} and Γ_{+} we obtain the two equations

$$\rho(\gamma^{\mu}p_{\mu}-mc)|\psi_{-},\tfrac{3}{2}\rangle = 0 \;, \qquad (2.31)$$

$$\rho(\gamma^{\mu}p_{\mu}-mc)|\psi_{+},\tfrac{3}{2}\rangle - 2i\tau_{0}\gamma^{\mu}p_{\mu}\gamma_{-}|\psi_{-},\tfrac{3}{2}\rangle = 0 \;. \qquad (2.32)$$

From (2.32) by operating with $(\gamma^{\mu}p_{\mu}-mc)^{-1}$ we obtain

$$\rho|\psi_{+},\tfrac{3}{2}\rangle = 2i\tau_{0}(p^{2}+mc\gamma^{\mu}p_{\mu})\gamma_{-}\left(\frac{1}{p^{2}-m^{2}c^{2}}\right)|\psi_{-},\tfrac{3}{2}\rangle \;,$$

which, because of (2.31), implies division by 0 (i.e. $p^2=m^2c^2$) unless we take $|\psi_{-},\tfrac{3}{2}\rangle = 0$ as the only solution of (2.31). Hence $|\psi_{+},\tfrac{3}{2}\rangle$ must also vanish. Therefore the wave equation (2.23) does not have non-zero $s=\tfrac{3}{2}$ solutions.

By projecting out with $\Gamma_{\pm}$ the wave equation (2.28) for $s=\tfrac{1}{2}$ splits up into the two equations

$$\rho(\gamma^{\mu}p_{\mu}-mc)|\mu\rangle - 3i\tau_{0}\gamma^{\mu}p_{\mu}\gamma_{-}|e\rangle = 0 \;. \qquad (2.33)$$

$$\rho(\gamma^{\mu}p_{\mu}-mc)|e\rangle + i\tau_{0}\gamma^{\mu}p_{\mu}\gamma_{-}|\mu\rangle = 0 \;, \qquad (2.34)$$

where

$$|\mu\rangle = |\psi_{-},\tfrac{1}{2}\rangle,\; |e\rangle = |\psi_{+},\tfrac{1}{2}\rangle \qquad (2.35)$$

and where the lepton wave function

$$|e\rangle+|\mu\rangle = \Gamma_{+}|\psi,\tfrac{1}{2}\rangle + \Gamma_{-}|\psi,\tfrac{1}{2}\rangle \;, \qquad (2.36)$$

obeys the wave equation

$$[\gamma^\mu p_\mu - mc(\gamma_- + a\gamma_+)]|\psi,\tfrac{1}{2}\rangle = 0 \quad , \qquad (2.37)$$

where

$$a = \frac{\rho^2(p^2-m^2c^2)}{\rho^2(p^2-m^2c^2)-3p^2} = \frac{\rho}{\rho \pm \sqrt{3}} \quad , \qquad (2.38)$$

and the + sign stands for the electron and the − sign for the muon. Thus the electron and muon belong to the eigenvalues +1 and −1 of Γ_5, respectively. The eigenfunctions $|e\rangle$ and $|\mu\rangle$ are, of course, orthogonal.

By setting $\rho=0$ in (2.33) and (2.34) we obtain the wave equations for ν_e and ν_μ and their anti-particles. If in (2.37) we had $a=1$ then it would be invariant under a parity transformation, which is possible only for $\rho=\infty$. This is equivalent to turning off the mass difference between μ and e. Setting $\rho=0$ in (2.37) again yields the wave equation for the neutrinos.

Now, on writing the equation (2.37) in the form

$$(\gamma_- + \tfrac{1}{a}\gamma_+)\gamma^\mu \partial_\mu |\psi,\tfrac{1}{2}\rangle - i\kappa|\psi,-\tfrac{1}{2}\rangle = 0 \quad ,$$

we obtain the conserved current density

$$J_\mu = \langle \bar\psi,\tfrac{1}{2}|(\gamma_- + \tfrac{1}{a}\gamma_+)\gamma_\mu|\psi,\tfrac{1}{2}\rangle \quad , \qquad (2.39)$$

where

$$\frac{1}{a} = 1 \pm \frac{\sqrt{3}}{\rho} \quad . \qquad (2.40)$$

For purely leptonic processes we can write down the weak currents

FUNDAMENTAL INTERACTIONS

$$\mathcal{L}_\mu = \langle\bar{\psi},\tfrac{1}{2}|(1 + \tfrac{\sqrt{3}}{\rho}\Gamma_5\gamma_+)\gamma_\mu|\nu\rangle , \qquad (2.41)$$

$$\mathcal{L}_\mu^\dagger = \langle\bar{\nu}|(1 + \tfrac{\sqrt{3}}{\rho}\Gamma_5\gamma_+)\gamma_\mu|\psi,\tfrac{1}{2}\rangle , \qquad (2.42)$$

which are of the form

$$\mathcal{L}_\mu = \langle\bar{e}|(1 + \tfrac{\sqrt{3}}{\rho}\gamma_+)\gamma_\mu|\nu_e\rangle + \langle\bar{\mu}|(1 - \tfrac{\sqrt{3}}{\rho}\gamma_+)\gamma_\mu|\nu_\mu\rangle , (2.43)$$

$$\mathcal{L}_\mu^\dagger = \langle\bar{\nu}_e|(1 + \tfrac{\sqrt{3}}{\rho}\gamma_+)\gamma_\mu|e\rangle + \langle\bar{\nu}_\mu|(1 - \tfrac{\sqrt{3}}{\rho}\gamma_+)\gamma_\mu|\mu\rangle . (2.44)$$

The interaction Lagrangian density is given by

$$\mathcal{L}^W = \tfrac{G}{\sqrt{2}}(\mathcal{L}_\mu^\dagger \mathcal{L}^\mu + \mathcal{L}^\mu \mathcal{L}_\mu^\dagger) , \qquad (2.45)$$

where

$$G = 1.435 \times 10^{-49} \text{ erg.cm}^3 .$$

If we adopt the W-meson model for the weak interactions then the corresponding equation for the W's in the case of the (μ,e) system is given by

$$\Gamma_{\mu\nu}\beta^\mu p^\nu \beta_- W + \rho(\beta^\mu p_\mu - mc)W = 0 . \qquad (2.46)$$

This equation yields the mass spectrum

$$\frac{m^2}{M^2} = (1 + \frac{\tau_o}{\rho}) \qquad (2.47)$$

where the parity matrix τ_o assumes the eigenvalues ± 1. The equation (2.46) for $\rho=0$ reduces to the photon equation.

Thus the positive and negative parity W-mesons are related by

$$\frac{m^2}{1+\frac{1}{\rho}} < m^2 < \frac{m^2}{1-\frac{1}{\rho}} \quad , \tag{2.48}$$

where

$$(M_{\overline{W}}^{\pm})^2 = \frac{m^2}{1+\frac{1}{\mp\rho}} \quad . \tag{2.49}$$

ON SL(6,C) GAUGE INVARIANCE

Abdus Salam

International Centre for Theoretical Physics,
Trieste, Italy, and Imperial College,
London, England

ABSTRACT

The Einstein-Cartan-Weyl theory of spin 2^+ particles is reviewed and its SL(2,C) gauge-invariant character brought out. The theory generalizes to describe a nonet of 2^+ particles when SL(2,C) invariance is extended to the spin and unitary-spin containing group SL(6,C).

I. INTRODUCTION

Wigner's spin-isotopic-spin symmetry SU(4) was extended to SU(6) in 1964 by Gürsey, Radicati and Sakita. Ever since, we have had a problem: (a) Is there something deep in this unification of intrinsic spin with internal symmetry? (b) Do there exist Lagrangians with both kinetic energy and interaction terms invariant under SL(6,C) or U(6,6)? (SL(6,C) is the relativistic generalization of SU(6) and U(6,6) the relativistic generalization of U(6) x U(6).) (c) Equivalently, do there exist <u>conserved</u> currents which close on the algebra of SL(6,C)? If there do, what is their physical significance?

A parallel problem was posed by the theory of spin 2^+ strongly interacting mesons. In 1970 Isham, Strathdee and Salam and, independently, Wess and Zumino had suggested that the 2^+ f^0 meson may be described by an equation similar to Einstein's with the field $f^{\mu\nu}$ replacing the graviton field $g^{\mu\nu}$. (One would of course also change the coupling parameter and supplement Einstein's equation with a mass term.) The problem with this suggestion was: how does one incorporate SU(3) (necessary to describe the known <u>nonet</u> of 2^+ particles) with the closely-knit space-time structure of Einstein's equation?

Clearly, both these problems are related to each other in that both require for their resolution a unification of internal symmetries with some sort of space-time structure.

I wish to report today some work that Isham, Strathdee and I have recently done in this direction. I shall essentially supply a critique of two notes we have written on the subject (Lettere al Nuovo Cimento <u>5</u>, 969 (1972) and ICTP, Trieste, preprint IC/72/155).

Briefly, what we have discovered is that Weyl had already shown in 1929 that the Einstein-Cartan gravitational Lagrangian possesses gauge invariance under the non-compact symmetry SL(2,C). When we rewrite the Einstein-Cartan-Weyl theory in a Dirac γ basis, <u>we find that this permits an instant generalization of SL(2,C) to SL(6,C) and, in particular, of Cartan's equation for</u>

SL(6,C) GAUGE INVARIANCE

spin-torsion, which generalizes to include what one may call internal-spin-torsion. One obtains thus a rather elegant unification of intrinsic spin with internal symmetry.

I am hoping very much that our approach may, at the least, make the beautiful geometrical ideas of Einstein, Weyl and Cartan accessible to particle physicists, though our motivation is to use these ideas for strong interaction physics. The literature on general relativity places so much emphasis on the GL(4,R) group of general coordinate transformations, that the other invariances of the theory, like SL(2,C) gauge invariance, so much more relevant to particle physicists' experience, tend to be ignored. We hope the balance gets somewhat redressed by our demonstration of the value of these other invariances of the Einstein-Cartan theory.

During the second part of my talk I shall be concerned with the problem of the particle spectrum given by the SL(6,C) gauge-invariant Lagrangian of Weyl type. In particular we must ensure (like Einstein and Weyl) that, considered as a classical Lagrangian, ours permits of the excitation of only the positive frequencies. Here we shall use again and again the ideas of spontaneous symmetry breaking and non-zero expectation values of tensor fields [$<f^{\mu\nu}> = \lambda \, \eta^{\mu\nu}$] - amazingly enough, essentially first introduced by Einstein (though not using this language) and now so much in prominence in particle physics.

II. GAUGING THE NON-COMPACT GROUP SL(2,C). EINSTEIN-CARTAN-WEYL THEORY

2.1 The twin ideas of gauge-invariant Lagrangians

and spontaneous symmetry breaking have played an important role recently in particle physics. As I have said before, both ideas go back fifty years to the work of Weyl and Einstein. Let us first take the gauge Lagrangians.

DEVELOPMENT OF IDEAS IN GAUGE THEORIES

		Gauge Group	Gauge particle(s)
1.	(1918); recognition by Weyl that the Maxwell Lagrangian is a gauge Lagrangian.	U(1)	1^- photon
2.	(1929); recognition by Weyl, Fock, Ivanenko that the Einstein-Cartan Lagrangian (1922) is a gauge Lagrangian	non-compact SL(2,C)	2^+ graviton
3.	(1954); generalization by Yang, Mills and Shaw of Weyl's 1918 theory.	SU(2)	1^- triplet ρ^+, ρ^0, ρ^-
4.	(1972); present generalization of Weyl's theory of 1929.	SL(6,C)	nonet of 2^+ particles $F, F', A_2 K^{**}$

Let us consider the Einstein-Cartan theory of gravity in Weyl's formulation. As is well known, Einstein's gravity theory works with a ten-component symmetric tensor field $g^{\mu\nu}(x)$. In order to describe gravitational interaction of spin-$\frac{1}{2}$ particles, Weyl introduced the so-called vierbein 16-component fields $L^{\mu a}(x)$, which bear to $g^{\mu\nu}(x)$ the same relation as Dirac's γ matrices bear to the unit matrix, i.e. $L^{\mu a}$'s are essentially a square root of $g^{\mu\nu}$, in the sense

SL(6,C) GAUGE INVARIANCE

$$\eta_{ab} L^{\mu a}(x) L^{\nu b}(x) = g^{\mu\nu}$$

where

$$\eta_{ab} = \begin{bmatrix} 1 & & & \\ & -1 & & \\ & & -1 & \\ & & & -1 \end{bmatrix}$$

In the sense of interpolating fields, both fields, $g^{\mu\nu}$ and $L^{\mu a}$, are at par; both could describe the graviton equally well.

Now, one of the curses of relativity theory is the multiplicity of indices which, in the cluster of formalism, successfully obscure the real heart of the basic ideas. In order to minimize indices, I shall use the Dirac γ basis to write my formulae. For SL(2,C) I shall need the four γ_a's and the six generators σ_{ab}'s, with commutation relations of the type:

$$[\sigma, \sigma] = i\,\sigma$$

$$[\gamma, \sigma] = i\,\gamma$$

In this Dirac basis the expression $L^\mu(x)$ will stand for the 4 x 4 matrix combination of Weyl fields $L^{\mu a}\gamma_a$. In particular, $g^{\mu\nu} = \frac{1}{4} \mathrm{Tr}(L^\mu L^\nu)$

2.2. Consider now a spinor $\psi(x)$ which, for the SL(2,C) index transformations, transforms as

$$\psi(x) \to \psi' = \Omega\psi(x),$$

where

$$\Omega = \exp i[\sigma_{ab}\varepsilon^{ab}] \quad .$$

Note that there is no x transformation implied at this stage. The only role the x coordinate will play will come about when we consider, in the standard gauge fashion, the parameters ε^{ab} to be functions of x. If $\varepsilon^{ab} = \varepsilon^{ab}(x)$, clearly

$$\partial_\mu \psi \neq \Omega \, \partial_\mu \psi(x) \quad .$$

The ordinary derivative does not transform in a simple manner. To "correct" this, introduce in the standard gauge fashion, the 24-component gauge field $B_\mu(x) = B_\mu^{ab} \sigma_{ab}$, ($B_\mu^{ab} = -B_\mu^{ba}$) – also called the "Weyl connection." Provided that B_μ transforms as

$$B_\mu \to \Omega B_\mu \Omega^{-1} - i \Omega \, \partial_\mu \Omega^{-1} \quad ,$$

the "Weyl covariant" derivative,

$$\nabla_\mu \psi = (\partial_\mu + iB_\mu) \psi \quad ,$$

transforms "correctly" in the same manner as ψ itself; i.e.

$$\nabla_\mu \psi \to \Omega(x) \nabla_\mu \psi \quad .$$

Also, since

$$L^\mu \to \Omega(x) L^\mu(x) \Omega^{-1}(x) \quad ,$$

we may define

$$\nabla_\mu L^\nu = \partial_\mu L^\nu + i[B_\mu, L^\nu] \quad ,$$

SL(6,C) GAUGE INVARIANCE

which will transform as:

$$\nabla_\mu L^\nu \to \Omega (\nabla_\mu L^\nu) \Omega^{-1} .$$

Finally, from the transformation law of B_μ, one can easily verify that the "covariant curl"

$$B_{\mu\nu} = \partial_\mu B_\nu - \partial_\nu B_\mu + i[B_\mu, B_\nu]$$

transforms as:

$$B_{\mu\nu} \to \Omega B_{\mu\nu} \Omega^{-1} .$$

We are now ready to write SL(2,C) invariant Lagrangians. Summarizing the transformations:

$$\psi \to \Omega \psi \quad , \quad \overline{\psi} \to \overline{\psi}\, \overline{\Omega} \quad (\overline{\Omega} = \Omega^{-1})$$

$$L^\mu \to \Omega L^\mu \Omega^{-1} \quad , \quad B_{\mu\nu} \to \Omega B_{\mu\nu} \Omega^{-1}$$

$$\nabla_\mu \psi \to \Omega \nabla_\mu \psi \quad , \quad \nabla_\mu \psi = (\partial_\mu + iB_\mu)\psi ,$$

we immediately see that

$$L_{matter} = i\overline{\psi}(L^\mu \nabla_\mu)\psi + m\overline{\psi}\psi \text{ is SL(2,C) gauge invariant,}$$

$$L_{Weyl} = -i \, \text{Tr}\, [L^\mu, L^\nu] B_{\mu\nu} \text{ is SL(2,C) gauge invariant .}$$

<u>The amazing fact is that the beautifully simple expression L_{Weyl} will turn out to be identical to the well-known Einstein Lagrangian for gravity, when no matter is present.</u> When spin-$\frac{1}{2}$ matter is present, $L_{Weyl} + L_{matter}$ gives Cartan's generalization (1922) of Einstein's theory. I shall demonstrate this equivalence in a

heuristic fashion presently. Before doing this, however, let us consider this Lagrangian and the field equations following from it in some more detail.

2.3. Consider

$$L = \text{Tr}\left(-i\,[L^\mu,L^\nu](\partial_\mu B_\nu - \partial_\nu B_\mu + i\,[B_\mu,B_\nu])\right) + i\bar\psi\,L^\mu(\partial_\mu + iB_\mu)\psi + m\bar\psi\psi + \text{h.c.} \qquad (2)$$

The field equations are:

$$\boxed{\begin{array}{ll} i\,[L^\mu, B_{\mu\nu}] = T_\nu & \text{Einstein's curvature equation} \qquad (3) \\ \nabla_\mu[L^\mu,L^\nu] = S^\nu & \text{Cartan's torsional equation} \qquad (4) \end{array}}$$

Here T_ν and S^ν are the matter stress-tensor density and the matter intrinsic-spin densities, respectively.

Operating on the second equation by ∇_ν and using the first equation we obtain

$$\boxed{\nabla_\nu S^\nu = i\,[L^\nu, T_\nu] \qquad \underline{\text{the tetrode identity}}.} \qquad (5)$$

Cartan's equation (4) written out in detail reads:

$$i\,[B_\mu,[L^\mu,L^\nu]] = S^\nu_{\text{matter}} - \partial_\mu[L^\mu,L^\nu] \qquad (6)$$

Essentially this equation tells us that B_μ can be solved in terms of $\partial_\mu L^\nu$, L^ν and matter spin density. When this solution for B_μ is substituted into Einstein's equation, we obtain a second-order equation for L^μ (or equivalently for the metric tensor $g^{\mu\nu}$) when there is no spin-$\frac{1}{2}$ matter present.

SL(6,C) GAUGE INVARIANCE

Now define the currents J^ν:

$$J^\nu = S^\nu_{\text{matter}} - i[B_\mu, [L^\mu, L^\nu]] \,. \qquad (7)$$

Clearly,

$$\partial_\nu J^\nu = \partial_\nu \partial_\mu [L^\mu, L^\nu] \equiv 0 \,. \qquad (8)$$

It is easy to show that this conserved set of currents J^ν are indeed the Noether currents for our Lagrangian and close on the algebra of SL(2,C) (see ICTP, Trieste, preprint IC/72/155,Sec.V).

2.4. Consider now the equivalence of the theory described above with Einstein's gravitational theory, when no matter is present. First note an algebraic identity due to Möller:

$$L = \text{Tr}[-i[L^\mu, L^\nu]B_{\mu\nu}] \equiv \text{Tr}\{(\nabla_\mu L^\nu)(\nabla_\nu L^\mu) - (\nabla_\mu L^\mu)(\nabla_\nu L^\nu)\}$$

$$+ \text{ a surface term} \qquad . \quad (9)$$

So far we have ignored space-time transformations of $L^\mu(x)$ and $B_\mu(x)$. Let us assume with Einstein that these transform as standard contra- and covariant quantities:

$$B_\mu(x) \to B'_\mu(x) = \frac{\partial x^\nu}{\partial x'^\mu} B_\nu(x) \qquad (10)$$

$$L^\mu(x) \to L'^\mu(x) = \frac{\partial x'^\mu}{\partial x^\nu} L^\nu(x) \,. \qquad (11)$$

The remarkable thing about the expression $-\text{Tr } i[L^\mu, L^\nu]B_{\mu\nu}$ is that it transforms like a scalar for the <u>general coordinate transformations</u>, in Einstein's sense, the reason for this being that $B_{\mu\nu}$ has the character of a "curl". We may now use

$$g^{\mu\nu} = \frac{1}{4} \text{Tr } L^\mu L^\nu \tag{12}$$

to raise the lower indices. To link with Cartan, we may define the Cartan covariant derivative (denoted with a double stroke $||$) which must take account both of the general coordinate transformations and the Weyl SL(2,C) transformations. Quite generally the linearity of this "connection" requires that:

$$L^\nu_{\ ||\mu} = \partial_\mu L^\nu + i[B_\mu, L^\nu] + \binom{\nu}{\mu\rho} L^\rho \ .$$

Here $\binom{\nu}{\mu\rho}$ is a generalized Christoffel-like (asymmetric) connection, for which we shall demand that the Cartan-derivative of L^μ (or equivalently of $g^{\mu\nu}$) vanishes. Thus $\binom{\nu}{\mu\rho}$ is defined by the relation:

$$0 = L^\nu_{\ ||\mu} = \nabla_\mu L^\nu + \binom{\nu}{\mu\rho} L^\rho \tag{13}$$

Using (13), clearly, the Möller form of Weyl's Lagrangian,

$$(\nabla_\mu L^\nu)(\nabla_\nu L^\mu) - (\nabla_\mu L^\mu)(\nabla_\nu L^\nu) \ ,$$

reduces to a form familiar from Einstein (when no matter is present):

$$L_{\text{Einstein}} = g^{\rho\rho'} \left(\binom{\mu}{\nu\rho} \binom{\nu}{\mu\rho'} - \binom{\mu}{\mu\rho} \binom{\nu}{\nu\rho'} \right) \ .$$

(For completeness, one should remark that one must divide L_{Einstein} (or L_{Weyl}) by the factor $\sqrt{-\det g^{\mu\nu}}$ =

$\sqrt{-\det(\frac{1}{4}\operatorname{Tr} L^\mu L^\nu)}$ in order that $\int L\, d^4x$ transforms as
a scalar. Since in what follows we do not wish to worry
about general coordinate transformations but only about
the Poincaré set of space-time transformations, so far
as strong interaction physics is concerned, this refine-
ment can for the present be ignored.)

Note, in passing, that the Cartan equation (4),

$$\nabla_\mu [L^\mu, L^\nu] = S^\nu \;,$$

reads, in terms of the generalized "Christoffel"
connection:

$$[L^\nu(^{\mu}_{\mu\rho}) - L^\mu(^{\nu}_{\mu\rho}), L^\rho] = S^\nu \;.$$

It is easy to see that the equation relates the anti-
symmetric part of $(^{\nu}_{\mu\rho})$ in μ,ρ indices, i.e. the torsion
tensor, to the spin density.

2.5. As will be seen later, when we generalize
$SL(2,C)$ to $SL(6,C)$, it is the Cartan equation and the
spin-density S^ν which get generalized to include not
only spin but also internal spin. But before we exhibit
this generalization, consider: what is the effect of
the spin-torsion terms of Cartan in gravitational
theory? Kibble has shown that, to the first order in
the Newtonian constant G_N, the gravitational potential
between the two spin-$\frac{1}{2}$ particles acquires an extra
repulsive contact term proportional to $G_N(\bar\psi\, \gamma_5\gamma_\mu\psi)^2$,
which, in the non-relativistic limit, reduces to a
repulsive contact potential proportional to the square
of the spin-density. The important point about this
repulsive contact potential is that it is gravitational

in origin and comes about on account of the <u>torsional</u> characteristics of space-time structure.

Following from this, recently Trautman has argued that the singularities of gravitational collapse and cosmology may be prevented by the direct influence of spin on the geometry of space-time, in virtue of Cartan's equation above. Trautman considers a universe filled with spinning dust, with spins all aligned along one direction - due presumably to the influence of some cosmic magnetic field. The Einstein and Cartan equations are compatible with a Robertson-Walker line element,

$$(ds)^2 = (dt)^2 - (R(t))^2 ((dx)^2 + (dy)^2 + (dz)^2) \ .$$

For small R, the spin density on the right-hand side of Cartan's equation plays the role of a "repulsive potential," which counteracts the universal "attractive" gravitational force. A universe consisting of 10^{80} neutrons would attain R_{min} of the order of 1 cm and collapse no further.

I am mentioning this because later, when we have generalized the Einstein-Cartan-Weyl Lagrangian to an SL(6,C) invariant form, we may find some speculative reasons why spin alignments and isotopic-spin alignments should occur together in regions of extreme spin-isotopic-spin density.

III. GENERALIZATION OF EINSTEIN-CARTAN-WEYL THEORY TO SL(6,C) GAUGE INVARIANCE

The generators of the SL(6,C) group are given by $\sigma_{ab} \lambda^i$, $\gamma_5 \lambda^i$, λ^i, while $\gamma_a \lambda^i$, $i\gamma_a \gamma_5 \lambda^i$ give the appro-

SL(6,C) GAUGE INVARIANCE

priate SL(6,C) generalization of the "ideal" γ_α.

Generalize L^μ and B_μ to contain (4 x 72) components each (rather than 4 x 4 components); thus

$$L^\mu = L^{\mu a i}\left(\gamma_a \frac{\lambda^i}{2}\right) + L^{\mu a i 5}\left(i\,\gamma_a \gamma_5 \frac{\lambda^i}{2}\right) \tag{14}$$

$$B_\mu = B_\mu^{abi}\left(\frac{\sigma_{ab}\lambda^i}{4}\right) + B_\mu^i \frac{\lambda^i}{2} + B_\mu^{5i}\left(\frac{\lambda^i}{2}\gamma_5\right), \tag{15}$$

and now adopt the <u>same</u> expression (2) as the Lagrangian for strong interactions exhibiting SL(6,C) gauge invariance. (Here λ^i are the nine 3 x 3 Gell-Mann U(3) matrices.) It is a triumph of the Dirac basis for the SL(2,C) case that the formalism carries over directly from SL(2,C) to SL(6,C).

Later we shall see that we need to add some more terms to the Lagrangian (2), particularly in order that the particles described by the Lagrangian possess mass. However, at this stage, remark that the Einstein's curvature equation (3), Cartan's torsional equation (4), the tetrode identity (5) and the definition of conserved currents (6) (which now close on the algebra of SL(6,C)) carry over directly without change from the SL(2,C) case, except that we are now dealing with 72-beins rather than vierbeins. Also we remark that if the internal symmetry group were not a unitary group, but some other variety of Lie group, the generalization of SL(2,C) to include internal symmetries might have presented difficulties.

IV. THE PARTICLE SPECTRUM

As I said earlier, so far as strong interaction physics is concerned, we shall not worry, for the present, about general coordinate transformations. The symmetry group to which we shall specialize has the structure of a semi-direct product

$$P \oslash SL(6,C) ,$$

where P denotes the Poincaré group. (The distinction of upper and lower indices is now trivial ($L^\mu = \eta^{\mu\nu} L_\nu$).)

Before considering the complicated $SL(6,C)$ case, let us examine the meaning of $SL(2,C)$ gauge invariance for the Einstein-Weyl-Cartan theory and introduce with Einstein the ideas of spontaneous symmetry breaking.

4.1. The Einstein-Weyl Lagrangian (and also the spin-$\frac{1}{2}$ Lagrangian in the limit m = 0) possesses no terms bilinear in field variables. If we assume with Einstein that

$$L^{\mu a}(x) = \eta^{\mu a} + \kappa \, \phi^{\mu a}(x) \tag{16}$$

where κ is the (strong gravity) coupling constant and $\phi^{\mu\alpha}(x)$ is the (quantized) field variable with zero expectation value, then

$$\langle L^\mu \rangle = \gamma^\mu . \tag{17}$$

The symmetry-breaking implied by $\langle L^{\mu a} \rangle = \eta^{\mu a}$ <u>provides a bridge, through an identification of Greek (μ) and Latin (a) indices, between the Poincaré transformations and the index transformations $SL(2,C)$</u>. If we now set

$$L^\mu = \gamma^\mu + \kappa \phi^\mu$$

in the Einstein-Weyl-Dirac Lagrangian, we do recover a

SL(6,C) GAUGE INVARIANCE

set of bilinear terms and with them a particle spectrum. In fact, symbolically, the structure of our Lagrangian now looks like:

$$\begin{aligned} L_{Weyl} &= L^2 \partial B + L^2 B^2 + \bar{\psi} L \partial \psi \\ &= \left\{ \phi \partial B + B^2 \right\} + \left\{ \phi^2 \partial B + \phi^2 B^2 \right\} \\ &\quad + \bar{\psi} \gamma \partial \psi \right\} + \left\{ \bar{\psi} \phi \partial \phi \right. \end{aligned}$$

Approximating L by its bilinears, we recover the field equations,

$$B = \partial \phi \quad , \quad \partial B = 0 \rightarrow \partial^2 \phi = 0 \; ;$$

telling us that we are dealing with a massless field ϕ (the graviton).

Before proceeding, let us add to L_{Weyl} further SL(2,C) gauge-invariant terms which give the particle mass, through the familiar intrinsic symmetry-breaking mechanism. Write

$$L_{mass} = \mathrm{Tr}(\beta_1 L^\mu L_\mu + \beta_2 (L^\mu L_\mu)^2 + \beta_3 (L^\mu L^\nu)(L_\mu L_\nu)) \; . \quad (18)$$

We shall see later that this term indeed gives rise to a mass M for 2^+ particles, provided $\beta_1 + 8\beta_2 - 4\beta_3 = 0$ and $\beta_1 = -(3M^2/2\kappa^2)$, $\beta_2 = -\beta_3 = (M^2/3\kappa^2)$ consistent with the spontaneous symmetry-breaking ansatz $\langle L^\mu \rangle = \gamma^\mu$, in a manner very familiar nowadays from the Higgs-Kibble theory. (Freund and Maheshwari have remarked that in a GL(4,R) theory, which is invariant for general coordinate transformations, $\mathrm{Tr}\, L^\mu L_\mu$ is a constant and we would not have been able to obtain mass from a symmetry-breaking formalism.)

4.2. Consider now the meaning of SL(2,C) gauge invariance. Infinitesimally,

$$L^\mu \to L^\mu + i[\epsilon^{ab} \sigma_{ab}, L^\mu] \quad .$$

<u>Clearly, the gauge transformation affects only the antisymmetric parts of $L^{\mu a}$</u> ($L^{[\mu a]} = \frac{1}{2}(L^{\mu a} - L^{a\mu})$) <u>infinitesimally</u>. In fact, SL(2,C) gauge invariance of the Lagrangian is simply the statement that the antisymmetric components of $L^{\mu a}$ <u>do not represent dynamical degrees of freedom</u> and we can specialize to a gauge where these can be set equal to zero. The bilinear part of the Weyl Lagrangian now reads:

$$L_{(2)} = -\frac{2}{\kappa^2} B_\mu^{\nu\alpha} (\phi_{\mu\alpha,\nu} - \eta_{\mu\nu}\phi_{\lambda\alpha,\lambda})$$

$$- \frac{1}{\kappa^2}(B_\mu^{\nu\alpha} B_\nu^{\mu\alpha} - B_\mu^{\mu\alpha} B_\nu^{\nu\alpha})$$

$$+ 8\beta_3 \phi_{\mu\nu}\phi_{\mu\nu} + 4(\beta_2 - \beta_3)\phi_{\mu\mu}\phi_{\nu\nu} \quad .$$

This is the well-known Pauli-Fierz Lagrangian where the only particle excitations correspond to those of a mass M and spin 2^+ particles, provided

$$\beta_1 = -\frac{3M^2}{2\kappa^2} \quad , \quad \beta_2 = -\beta_3 = \frac{M^2}{8\kappa^2} \quad . \qquad (19)$$

V. PARTICLE SPECTRA FOR SL(6,C) AND SL(6,C) × SL(6,C) GAUGE LAGRANGIANS

5.1. Consider now the Weyl Lagrangian,

$$L = \text{Tr } i [L^\mu, L^\nu] B_{\mu\nu} + L_{mass} \quad \text{(given by Eq. (18))}, \qquad (20)$$

SL(6,C) GAUGE INVARIANCE

generalized to SL(6,C) and once again consider the bilinears obtained by setting

$$L^\mu = \gamma^\mu + \kappa\phi^\mu .$$

As one can see from (14) and (15), in addition to nonets of conjugate fields $L^{\mu a(k)}$ and $B_\mu^{ab(k)}$, the SL(6,C) theory needs the introduction of the following extra fields: $L^{\mu a 5(k)}$, $B_\mu^{(k)}$, $B_\mu^{5(k)}$. If one examines the bilinear part of (20), one finds that the conjugate set $L^{\mu a(k)}$, $B_\mu^{ab(k)}$ correctly gives the propagation of a massive 2^+ nonet. The extra fields $L^{\mu a 5(k)}$, $B_\mu^{(k)}$ and $B_\mu^{5(k)}$, however, make their appearance only in one place among the kinetic energy bilinears in a term which reads:

$$L^{\mu\nu 5(k)} (B_{\nu,\mu}^{5(k)} - B_{\mu,\nu}^{5(k)}) .$$

These fields do appear in trilinear and quadrilinear parts of the Lagrangian, but the bilinear terms give no clue as to their propagation character. This implies that <u>either</u> we should devise methods by which we can infer the particle spectrum corresponding to these fields from the trilinear and quadrilinear parts of the Lagrangian, <u>or</u> we should supplement the Weyl Lagrangian (20) by additional SL(2,C) invariant terms. These should be such as to give new sets of bilinears which should guarantee the (positive metric) propagation properties for the extra fields.

In the following I shall illustrate both approaches.

5.2. Before going on to consider the problem posed above, let me return for a moment to Weyl's SL(2,C) gauge-invariant Lagrangian and try to bring out the significance of its gauge invariance in a slightly different manner. The remarks I shall make will be relevant to the problem of propagation of the extra fields $L^{\mu a 5}$, B_μ and B_μ^5.

Given a 16-component field quantity L^μ (with a non-zero expectation value $<L^\mu> = \gamma^\mu$), one has a mathematical theorem - the so-called "polar decomposition theorem" - which states that one can write L^μ uniquely in the form:

$$L^\mu = S \ell^\mu S^{-1}, \qquad (21)$$

where ℓ^μ is symmetric in the sense $\ell^{\mu a} = \ell^{a\mu}$ and S has the form:

$$S = \exp iP \; ; \; P = P^{ab} \sigma_{ab} .$$

To prove this result, set up an iteration system,

$$L^\mu = \gamma^\mu + L_1^\mu + L_2^\mu + \ldots, \quad \ell^\mu = \gamma^\mu + \ell_1^\mu + \ell_2^\mu + \ldots,$$

$$P = P_1 + P_2 + \ldots$$

The relation (21) is equivalent to the set of equations

$$\ell_1^\mu = L_1^\mu - i [P_1, \gamma^\mu]$$

$$\ell_2^\mu = L_2^\mu - i [P_1, L_1^\mu] - i [P_2, \gamma^\mu] . \qquad (22)$$

Clearly, the postulated symmetry of ℓ_1^μ implies that ℓ_1^μ and P_1 are respectively given by the symmetrical and antisymmetrical parts of L_1^μ and so on. Clearly, also in terms of the "polar decomposition" above, we can understand the transformation

SL(6,C) GAUGE INVARIANCE

$$L^\mu \to \Omega L^\mu \Omega^{-1} \qquad (23)$$

as equivalent to the transformation

$$S \to \Omega S, \qquad (24)$$

with ℓ^μ as a scalar quantity so far as the SL(2,C) transformations are concerned.

Given S as a functional of L^μ, define now a lower-case quantity b_μ through the relation

$$B_\mu = S b_\mu S^{-1} - i S \partial_\mu S^{-1} \qquad (25)$$

Combined with the transformation law for B_μ, viz.,

$$B_\mu \to \Omega B_\mu \Omega^{-1} - i \Omega \partial_\mu \Omega^{-1}, \qquad (26)$$

the relation (25) guarantees that b_μ is a scalar. The invariance of

$$L = \text{Tr } i \, [L^\mu, L^\nu] B_{\mu\nu} \qquad (27)$$

is now simply the statement that L identically equals

$$L = \text{Tr } i \, [\ell^\mu, \ell^\nu] b_{\mu\nu}, \qquad (28)$$

and that there is no S-dependence of the Lagrangian.

5.3. Let me now return to the problem posed earlier: the problem of propagation of the fields $L^{\mu a\,5}(k)$, $B_\mu^{(k)}$ and $B_\mu^{5(k)}$.

So far as the variables $L^{\mu a\,5}(k)$ are concerned, it appears that a simple "classical completion" of the

Lagrangian is provided by an extension of the gauge group from $SL(6,C)$ to $SL(6,C) \times SL(6,C)$. The details are worked out in Sec. IV of ICTP, Trieste, preprint IC/72/155. Here I wish to illustrate one important new idea which we had to introduce to cut down the multiplicity of fields and to guarantee a positive metric for the particles described by $L^{\mu a 5}(k)$. This is the idea of $(SL(6,C) \times SL(6,C))$ <u>covariant constraints</u>.

Consider gauge transformations of the type

$$\Omega = \exp(i\beta) \exp(\gamma), \qquad (29)$$

where β and γ contain 72 parameters each, corresponding to $SL(6,C) \times SL(6,C)$. The theory would possess two gauge fields now:

$$(B + iC)_\mu \to \Omega(B + iC)_\mu \Omega^{-1} - i\Omega \partial_\mu \Omega^{-1}$$

$$(B - iC)_\mu \to (\overline{\Omega})^{-1}(B - iC)_\mu \overline{\Omega} - i\overline{\Omega}^{-1} \partial_\mu \overline{\Omega} \qquad (30)$$

So far, everything is straightforward. But now comes the subtle new feature of the theory. In general we should work with two distinct L-type fields, with the transformation characters,

$$L_1^\mu \to \Omega L_1^\mu \overline{\Omega} \qquad (31)$$

$$L_2^\mu \to \overline{\Omega}^{-1} L_2^\mu \Omega^{-1} \qquad (32)$$

In order, however, to reduce the independent degrees of freedom, we can tie these two fields together, defining one as a non-linear functional of the other, consistent with the transformation laws (31) and (32). To see this, assume we are given L_1^μ. One can show that a "polar decomposition" exists which permits us to write

$$L_1^\mu = S[\ell^{\mu a(k)} \gamma_a \lambda^{(k)} + \ell^{\mu a5(k)} i \gamma_a \gamma_5 \lambda^{(k)}]\bar{S} \, , \quad (33)$$

where S has the form

$$S = \exp iP \exp Q \, , \quad (34)$$

$\ell^{\mu a(k)} = \ell^{a\mu(k)}$ and $\ell^{\mu a5(k)} = \ell^{a\mu 5(k)}$.

Just as the SL(2,C) case, it is easy to see that (31) and (33) are consistent with the transformations

$$S \to \Omega S \, , \quad (35)$$

and with the symmetric quantities $\ell^{\mu a}$ and $\ell^{\mu a5}$ transforming as scalars.

<u>Construct</u> now the field quantity

$$\bar{S}^{-1}[\ell^{\mu a(k)} \gamma_a \lambda^{(k)} - \ell^{\mu a5(k)} i \gamma_a \gamma_5 \lambda^{(k)}]S^{-1} \, . \quad (36)$$

Clearly this field is a non-linear functional of L_1^μ. The important point is that (on account of the crucial minus sign in front of $\ell^{\mu a5(k)}$ in (36)) it provides a representation of $L_{(2)}^\mu$ with the correct transformation law (32).

We are now in a position to write a simple SL(6,C) × SL(6,C) generalization of Weyl's Lagrangian. Consider

$$L = \text{Tr } i L_1^\mu L_2^\nu (B_{\mu\nu} + i C_{\mu\nu}) + \text{h.c.} + L_{mass} \, . \quad (37)$$

The gauge invariance of the Lagrangian implies that the only (physical) fields which occur in this expression are the fields $\ell^{\mu a(k)}$, $\ell^{\mu a5(k)}$ and the fields b_μ and c_μ defined by the relation

$$B_\mu + i C_\mu = S(b + ic)_\mu S^{-1} - i S \partial_\mu S^{-1} . \qquad (38)$$

It is easy to verify that the bilinears obtained from (37) by the spontaneous symmetry-breaking ansatz describe the propagation of two nonets (both with positive definite metrics): a 2^+ nonet, described by the fields $\ell^{\mu a(k)}$ and $b_\mu^{ab(k)}$, and a 2^- nonet, described by $\ell^{\mu a5(k)}$ and $c_\mu^{ab(k)}$. There are no bilinear terms for the fields b_μ, c_μ, b_μ^5 and c_μ^5.

5.4. Finally, now, we are confronted with the problem of the propagation character of the fields $b_\mu^{(k)}$, $b_\mu^{5(k)}$, $c_\mu^{(k)}$ and $c_\mu^{5(k)}$, which occur only among the tri-linear and quadrilinear terms in the SL(6,C) x SL(6,C) Lagrangian proposed in (37).

In the paper IC/72/155 I have been referring to, we have suggested the addition of one further <u>gauge-invariant</u> expression to the Weyl form of the Lagrangian which, so far as the bilinears are concerned, has the characteristic of adding mass-like terms $(b_\mu^k)^2$, $(c_\mu^k)^2$, etc. Thus our final Lagrangian is (37) plus <u>one</u> new gauge-invariant term to cope with the problem of the fields b_μ^k, c_μ^k,... As a consequence of this new term, the field equations express these fields as algebraic functions of the dynamical fields $\ell^{a\mu}$, $\ell^{\mu a5}$, b_μ^{ab}, and c_μ^{ab}. Today I should like to discuss a different procedure which I believe is more general and likely, in the long run, to be more important for dealing with degrees of freedom which make their appearance only among trilinear and quadrilinear terms of quantum

SL(6,C) GAUGE INVARIANCE

Lagrangians. We call this procedure "quantum completion."

Consider as an illustration the field $b_\mu^{(k)}$. The SL(6,C) x SL(6,C) Weyl Lagrangian (37) contains, among the trilinear terms, the expression

$$[b_{\nu,\mu} - b_{\mu,\nu} + b_\mu \times b_\nu] \cdot [\ell^{\mu\alpha} \times \ell^{\nu\alpha} - \ell^{\mu\alpha 5} \times \ell^{\nu\alpha 5}] .$$

(Here $\ell^{\mu\alpha} \times \ell^{\nu\alpha}$ is short for $f^{ijk} \ell^{\mu\alpha(j)} \ell^{\nu\alpha(k)}$.) In a classical sense, this Lagrangian does not tell us much about the propagation of the b_μ field. In a quantum sense, however, a second-order iteration of this term will give rise to a loop diagram:

$$\underline{\partial_\mu b_\nu - \partial_\nu b_\mu} \bigcirc \underline{\partial_\nu b_\mu - \partial_\mu b_\nu}$$

with $\ell^{\mu\alpha}$ and $\ell^{\nu\alpha}$ in the loop.

Since the propagators of the ℓ fields are known, this diagram can be computed, and immediately leads to an effective Lagrangian for the b_μ field. We have actually computed the loop diagram; its leading (most divergent) contribution is of the form:

$$L_{eff} = - [\partial_\mu b_\nu - \partial_\nu b_\mu + b_\mu \times b_\nu]^2 \frac{\delta^4(0)}{M^4}$$

with the <u>correct</u> metric for the positive frequency propagation of the b_μ field (provided we assume that $\delta^4(0) > 0$ and, in some renormalization sense, represents a finite number).

This idea of "quantum completion" is not new. It is closely linked with the old idea of using conditions like $Z = Z(\kappa^2, M^2) = 0$ to give propagation character to composite fields (which in fact the fields b_μ^k, c_μ^k,.....

etc., really are) since the Lagrangian, as it stands, gives no propagation equations for these. In fact, Sacharov and Zeldovitch and Durr and Heisenberg have gone so far as to suggest that one may recover the Maxwell field equations by starting with the Lagrangian

$$L = -\bar{\psi}\gamma(\partial + ieA)\psi - m\bar{\psi}\psi ,$$

with propagation terms for the ψ field only. One recovers the propagation character for the photon by considering the loop:

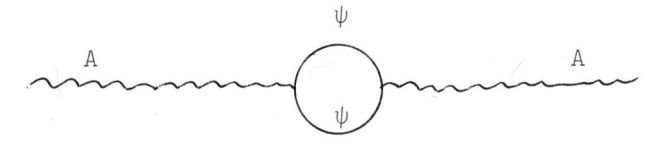

which, for special values of e determined by $Z(e^2) = 0$, would give rise to the effective Maxwell term $F_{\mu\nu}F_{\mu\nu}$. We are not going as far as Sacharov and Zeldovitch, since we have a simpler problem. Our Lagrangian already contains a first-order derivative for the b_μ field. A second-order iteration gives rise to an effective Lagrangian guaranteeing the propagation of b_μ apparently for all values of the coupling parameter.

To summarize, one may suggest (as an alternative to adding more terms to (37)), using the Weyl Lagrangian as it stands, to give a propagation character to a 2^+ and a 2^- nonet. In ensuring that both nonets possess positive definite metrics, we employed a non-linear realization of the L_2^μ field in terms of L_1^μ. The spin-1 fields B_μ, C_μ,... may acquire a propagation character through the process of "quantum completion." Their effective coupling parameters, however, will have no definite relation to the coupling parameters of the 2^+

and 2^- particles. If one does not like the idea of a "quantum completion," there is always the possibility of adding to the Weyl Lagrangian one extra "classical" term, which guarantees that these fields can be eliminated from the theory as algebraic constraints. Of course the "classical completion" and the "quantum completion" give rise to different theories. Before closing with this part of my talk, I wish to make two remarks. One is that Gürsey (<u>Contemporary Physics</u> (IAEA, Vienna 1969) p. 211) has discussed an SL(6,C) invariant - but not an SL(6,C) <u>gauge</u>-invariant theory, where, in our notation, a field L^μ is introduced through the definition $L^\mu = S \gamma^\mu S^{-1}$ with $S = \exp i P$. The 72-fields P are the basic fields and not just gauge fields as in our formulation. There are no B_μ fields, no Einstein-like Lagrangian and naturally no spin-2 gauge particles. The fields P must be zero-mass Goldstone fields in the symmetry limit. It seems to us that the relationship between Gürsey's and our theory is roughly the same as that between a (non-linearly realized) Goldstone theory and a Higgs-Kibble type of gauge theory.

The second point I wish to make is that, in our opinion, it is not one particular Lagrangian versus another which is likely to be important in strong-interaction physics in the long run. Arguing along with Gell-Mann, one must learn to "abstract" the basic truth from the outer wrappings of formalism. We believe the truth in this instance lies in the deep geometrical ideas of Cartan regarding torsion and its connection not only with spin but also with internal symmetries.

VI. TOWARDS SU(6)

I started this lecture by wanting to solve two problems: i) to find a possible origin for spin-internal-spin combination and thereby motivate SU(6); ii) to find an elegant generalization of Einstein's Lagrangian, describing spin 2^+ particles, so as to include SU(3).

What we have accomplished is to generalize Cartan's geometrical notion of torsion to include internal symmetries. We have also succeeded in finding an elegant theory of 2^+ (and 2^-) nonets. But have we succeeded in recovering SU(6)?

Two years ago, lecturing here at Miami, I remarked on Kerr's <u>exact</u> solution to Einstein's equation; the remark was that in the Kerr solution for a charged spinning particle (mass m, charge Q, spin J), the charge Q always occurs in the combination

$$Q^2 + \frac{J^2}{m^2 G_N} \quad .$$

I conjectured that once SU(2) or SU(3) is incorporated into the structure of the Einstein equation, Q^2 will generalize to an expression like I(I + 1), where I is the isotopic spin. I suggested that SU(2) or SU(3) containing Einstein-like equations, when solved <u>exactly</u> (as for the Kerr case), may provide the dynamical basis for the emergence of SU(4)- or SU(6)-like combinations,

$$I(I + 1) + \frac{J(J + 1)}{m^2 G_f} \quad ,$$

where G_f is the strong-gravity coupling parameter which

replaces the newtonian constant G_N. The situation with
SU(6) would then be similar to the situation for the
hydrogen atom, where the observed SO(4) symmetry of
hydrogen energy levels is a dynamical (and unexpected)
consequence of the 1/r potential. Now that we have
successfully incorporated SU(3) (or indeed SU(3) x SU(3))
into the Einstein-Cartan-Weyl-like equations, with (a
non-linearly realized) SL(6,C) x SL(6,C) as our starting point and with an algebra of conserved currents, it
seems eminently reasonable that SU(6) symmetry (with
its attendant manifestations in terms of collinear and
coplanar subgroups) should emerge dynamically from
different approximations of the theory we have constructed. So far as matter is concerned, presumably we would
consider writing the quark-anti-quark Lagrangian describing the 35-plet of SU(3) in the generalized Bergmann-
Wigner form

$$L = \frac{i}{2} \text{Tr} \, (\Phi[L^\mu, \nabla_\mu \Phi] - m\Phi\Phi) \quad,$$

where Φ is the second-rank multi-spinor. Likewise for
the third-rank multi-spinor describing the 56-plet of
the baryons. In addition we may have direct SL(6,C)-
invariant couplings of the type $\text{Tr} \, \Phi\Phi\Phi$. The 2^+ (and 2^-)
nonets described by the fields $L^\mu(x)$ will act as gluons.
(As stated earlier, these particles constitute incomplete multiplets of the old quark-based phenomenological
SU(6). The situation for these multiplets is presumably
completely analogous to <u>non-linearly</u> realized chiral
theories, which also display incomplete multiplets of
the larger symmetry group.) The true dynamics will be
a complicated interplay of exactly symmetric vertices,
like $\text{Tr} \, \Phi\Phi\Phi$, of the spontaneous symmetry-breaking

mechanism $\langle L^\mu \rangle = \gamma^\mu$ and of the covariant constraints, like those typified in (36).

One final and very speculative remark. In Sec. 2.5 we considered Trautman's inclusion of Cartan's torsional effects of spin for the problem of gravitational collapse. Trautman needed an alignment of spins to build up a finite spin density. If one considers the innermost core of a neutron star, with densities in excess of 10^{15} gm/cm^3 and with perhaps 10^{19} neutrons in the core region, it seems likely that the short-range order due to F-gravitational effects may be as important as the long-range order due to newtonian gravity (on account of G_F/G_N being of the order of 10^{38}). Is it conceivable that the spin-isotopic-spin independence of SL(6,C) invariant F-gravitational forces makes for a spin alignment of the neutrons in the core region, simply because a core made purely of neutrons must possess a perfect alignment of isotopic spins? Is it conceivable that high concentrations of isotopic-spin density (represented, for example, by very dense neutron stars) play, through the marriage of spin-isotopic-spin-torsional effects, a role in cosmology?

GEOMETRY AND PHYSICS OF THE ELEMENTARY PARTICLES

D. I. Blokhintsev

Joint Institute for Nuclear Research

Dubna, U. S. S. R.

My report is devoted to the problem of geometry in the physics of elementary particles.

1. ORDERING OF EVENTS IN MACROSCOPIC PHYSICS

I think it would not be an overstatement to say that theoretical physics starts with an ordering of events. This is the basis of all foundations. To perform this ordering each point-like event P is assigned a set of four numbers $x = x_0, x_1, x_2, x_3$, the coordinates of this event. If an event requires more than one set of four numbers it is not a point-like event. The operation of ordering is called arithmetization.

The way of performing arithmetization of events contains an essential element of agreement. The present agreement is based on the assumptions of i) universality of the constancy of the velocity of light and ii) existence of a uniformly running clock.* These assumptions

---

* One usually means a "light" clock, which is a light

are the bases of arithmetization of point-like events, accepted in the theory of relativity. For the coordinates of the point x to have physical sense, the existence of a real object, which might be a representative of a point-like event $P(x)$ at the coordinate x is necessary. I believe that in classical physics, to the concept of a point-like event $P(x)$, there corresponds the concept of a material point - an object of finite mass m and infinitely small dimensions $a \to 0$.

Within the framework of special relativity I don't see any logical contradictions between the Einstein method of arithmetization and the mechanics of material points. Therefore, in special relativity, material points may be thought of as objects which realize physically a point event $P(x)$.

2. RESTRICTIONS IMPOSED BY GRAVITATION

Let us consider a material point with mass m and dimensions $\underline{a}$. We favor the point with infinitely small dimensions ($a \to 0$) as the representative of a point-like event. However, if the dimensions of the point become less than its gravitational radius

$$a_g = \frac{2km}{c^3} \qquad (1)$$

(here $k = 6.7 \times 10^{-8}$ cm^3/g sec^2 is the gravitational constant, c is the velocity of light) then the metric

---

pulse reflecting periodically between two closely spaced mirrors. Then the assumption ii) is equivalent to the assumption of the existence of rigid standard of spatial separation between mirrors.

"inside" the point becomes nonstationary, and a collapse
ensues. At the same time, no signal sent from "inside"
the point can reach the external observer and, consequently, there is no way of getting information concerning the ordering of events "inside" the domain $r < a_g$.
Thus, with the given mass of the point m and the decrease of its dimensions, there comes a restriction on
the accuracy of measuring the coordinates of a point-like
event,

$$\Delta x \stackrel{\sim}{=} a = a_g \qquad (2)$$

In addition, the density of the matter ρ_g "inside" the
point is equal to or greater than

$$\rho_g = \frac{3}{4\pi} \left(\frac{c^2}{2k}\right)^3 \frac{1}{m^2}, \qquad (3)$$

and can go beyond the limits we are aware of from elementary particle physics.

Restrictions of another kind can come from the stochastic gravitational fields. The turbulent motion of
matter leads inevitably to the metric with a stochastic
metric tensor. If the fluctuations of this tensor are
not small then the ordering in the $R_4(x)$ space becomes
also essentially stochastic.

Spaces with a stochastic metric have been considered in the mathematical literature from an axio-
matic point of view. However, the spaces treated are
those with positive-definite metric. An extension of
axiomatization to stochastic spaces of the Minkowsky
type is still an open problem. Some of the related
problems have been discussed by the author.

The main question to be answered by physicists rather than by mathematicians is related to the indication of a physical method of ordering. Are we approaching here the limits of applicability of the concept of ordering?

Problems associated with the metric at large fluctuations and extremely large densities appear to become of foremost importance in the analysis of the earlier stages of the "Big Bang." We know nowadays that matter is governed by definite laws and that there exist symmetries. But we have no reason to assert that these laws and symmetries are time independent. It is quite possible that the vacuum and the world of elementary particles of which we are aware are nothing more than one of the possible ways of evolution of the Universe, chosen as a result of competition of various possibilities. However, at the present stage of our knowledge we do not have sufficient information to be able to discuss this facet of the problem in more detail.

3. POINT-LIKE EVENTS IN THE MICROWORLD

Now we pass to the world of elementary particles. Modern quantum field theory which describes the behavior of elementary particles is based on the condition of locality

$$[\phi(x), \phi(y)] = \mathcal{D}(x - y) \quad (4)$$

$$\mathcal{D}(x - y) = 0, \ (x - y)^2 < 0 \quad . \quad (5)$$

Here $\phi(x)$ is the field operator at a point x; $\phi(y)$ is the operator of the same field at a point y; $[A, B]$ is the commutator of the operators A and B.* The condition (4) is the causality principle and it implies the independence of the fields if the points x and y are separated by a space-like interval $(x - y)^2 < 0$. In other words, an arbitrary variation of the field at the point y cannot affect the field at the point x since a signal of velocity v < c cannot in this case reach the point x and vice versa.

In the condition of locality (5), the coordinates of the points x and y are assumed to be determined precisely. This assumption is equivalent to supposing the existence of point-like events $P(x)$, $P(y)$, ... We must analyze to what extent this assumption is consistent with the framework of the same local theory.

Elementary particles, the analogs of material points in classical physics, are natural candidates to play the role of representatives of point-like events. However, this analogy is inappropriate, due to a number of particular features dictated by the laws of quantum physics. The main point is that elementary particles cannot be localized in space and time more exactly than in the limit $\Delta x > \hbar/mc$.

In fact, in case of the Bose-Einstein particles, the density obeying the conservation law is equal to

$$\rho(\vec{x},t) = \phi^*(\vec{x},t)\hat{\Omega}\phi(\vec{x},t), \qquad (6)$$

where $\hat{\Omega} = (m^2 - \nabla^2)^{1/2}$ is the frequency operator. This density is positive-definite only in a nonrelativistic

---

*Or the anticommutator if the field is a spinor field We write explicitly the condition for the scalar field.

domain when $|\vec{v}| \ll m$. In case of the Fermi-Dirac particles, as is known, there exists a conserved density:

$$\rho(\vec{x},t) = \psi^*(\vec{x},t)\psi(\vec{x},t) . \qquad (7)$$

However, these are field equations; it is impossible to create a one-particle wave packet with $\Delta x < \hbar/mc$, and thus we see that in the microworld there are no objects which would serve as models of the point-like event $P(x)$.

In classical physics any material point can be considered not only as the realization of a point-like event but also as a body of reference ("Bezugkörper") which fixes the frame of reference. In the world of elementary particles this is impossible.

If we choose as a body of reference an elementary particle of rest mass m, then it is not difficult to show that the coordinates $\bar{x}$, $\bar{t}$ in this system of reference $\bar{\Sigma}$ will not be c-numbers, but q-numbers (operators). Between these operators there is a commutation relation:

$$[\bar{x},\bar{t}] = i\frac{\hbar}{mc} (x - \hat{v}t) \qquad (8)$$

where $\hat{v}$ is the three dimensional velocity of a particle of mass m in system Σ.

Thus, elementary particles of finite mass can be utilized neither as objects by means of which one marks points in the $R_4(x)$ space nor as bodies of reference. On the other hand, experimental facts point out that the predictions of the local field theory based on microscopic causality[10] are satisfied up to distance of the order of 10^{-15} cm. It might thus be supposed that there exist elementary particles with a mass much heavier than the mass of a nucleon, for which $\Delta x = \frac{\hbar}{m_p c} = 2 \times 10^{-14}$ cm.

Following this idea we come to the conclusion that <u>local theory assumes implicitly the existence of arbitrarily heavy elementary particles ($m \to \infty$)</u>. Under this assumption the contradiction between the use of the notion of arbitrary exact coordinates in the space $R_4(x)$ and the absence of objects playing the role of point-like events would be eliminated.

The limitation of the particle masses, from the above argument, to a certain limit $m < M$ would imply a principal limitation of applicability of the local theory for distances of the order of $\Delta x = \hbar/mc$. The requirements of an ideal experiment concerning the marking of space-time points are opposite in classical and quantum physics.

Now I consider possible reasons for the existence of the upper limit of the elementary particle mass.

4. THE RESTRICTIONS ON THE ELEMENTARY PARTICLE MASS BY GRAVITATIONAL CONSIDERATIONS

Just as in macroscopic physics, limitations on the elementary particle mass can come to the microworld from gravitation. It is not difficult to show that the fluctuation $\Delta g_{\mu\nu}$ of the metric tensor $g_{\mu\nu}$ is of the order

$$\Delta g \simeq \frac{a_g}{a}, \qquad (9)$$

where a_g is the gravitational radius of fluctuation and a is its linear dimension; when $a_g = a$ these fluctuations become very large (>1). At the same time $a_g = a$ is a collapse condition. Using this condition, in the case of an elementary particle with the mass m, we get

$$m = M_g = \frac{\hbar}{\Lambda_g c} \quad . \tag{10}$$

Here

$$\Lambda_g = \sqrt{\frac{8\pi k \hbar}{c^3}} = 0.82 \times 10^{-32} \text{cm} \tag{11}$$

is the known length expressed in terms of the gravitational constant k and the Planck constant $\hbar$. Thus the length defines the maximum mass of a particle obeying the laws of quantum theory. Such a particle was called a <u>maximon</u> by M. Markov.

Quite different conclusions can be drawn concerning the role of gravitation in the world of elementary particles, depending on to what frequencies Ω the spectrum of vacuum fluctuations is really limited. According to contemporary theory, it is almost uniformly distributed over all frequencies and the high frequencies are expected to give an infinitely large contribution to gravitation. If in the future, for some reason, it should turn out that the possible frequencies in the microworld are limited by an "elementary" length, then gravitational effects will become unimportant. In the opposite case they will play the fundamental role in the microworld, but they will be represented by their quantum interpretation.

Predictions and hopes based on classical calculations of gravitation will be "majorized" by quantum effects. At the same time the "classic" average metric will no longer be useful in these cases and a situation like the one mentioned in Section 2 will arise: the notion of interval between events as well as the idea itself about a possible ordering of events in $R_4(p)$ space becomes more than doubtful. Here we approach the edge of an abyss without being sure if it is not too early to peep into it.

5. THE PECULIARITIES OF WEAK INTERACTIONS

Now I consider other possible restrictions on local theory. Among various competitors the weak interaction lays claim to this role, because it has a universality similar to gravity.

We distinguish three modes of interactions: strong, electromagnetic and weak. Let us compare their behavior at high energies, using a criterion which I suggested several years ago. According to it, an interaction is assumed to be strong if in the course of the interaction the density of the kinetic energy ε_k is much smaller than the absolute density of the energy of the interaction W:

$$\varepsilon_k \ll |W| . \qquad (12)$$

Using this criterion we can show, that in case of nucleon-nucleon and pion-nucleon collisions, the interaction is strong at all energies if $g^2/\hbar c > 1$ (g is the strong coupling constant). On the contrary, according to the same criterion the electromagnetic interaction is always weak (as $\frac{e^2}{\hbar c} = 1/137$).

As for a weak interaction, it becomes strong at an energy

$$E_f = \frac{\hbar c}{\Lambda_f} = 300 \text{ GeV}, \tag{13}$$

where

$$\Lambda_f = \sqrt{\frac{G_f}{\hbar c}} = 0.66 \times 10^{-16} \text{ cm} \tag{16}$$

is the length characteristic of a weak interaction; G_f is the Fermi constant.

If the increase in strength of the weak interaction with energy is confirmed, then this will result in grave consequences for the theory.

Returning to the main theme of the report we can point out two such consequences.

6. WEAK MAXIMON

The ratio of the decay constant Γ to the hadron mass M under a weak decay of the kind

$$M \to m + l + \bar{\nu}$$

increases with the energy (here m is the nucleon mass, l is the lepton, and $\bar{\nu}$ stands for an antineutrino). The decay constant Γ for this process, when M>>m, is

$$\frac{\Gamma}{m} = \frac{1}{4\pi^3} G_f^2 M^4 N, \tag{15}$$

where $G_f = 1 \times 10^{-5}/m_p^2$ and N is the number of channels for different decay modes (which may be large). It is seen from this formula that for a hadron mass

$$M > M_f = \frac{\hbar}{\Lambda_f c}, \tag{16}$$

GEOMETRY AND PHYSICS

the decay constant Γ becomes comparable with the hadron mass M and the hadron can not really be considered an elementary particle since it may not be assigned a definite mass. It is advisable to refer to a particle with mass M_f as a <u>weak maximon</u>. As is seen from eqs. (16) and (10), this restriction begins to work earlier than that imposed by gravitation, since $M_f < M_g$. At the same time the restriction on local theory must then occur much earlier than is expected from the assumption about the existence of a gravitational maximon M_g.

7. THE "BLACKNESS" OF PARTICLES AND LOCALITY

The elementary particle can be thought of as a certain medium which is described by the creation and annihilation of virtual particles.

It is natural to raise the question about the condition of propagation of a metric signal in such a peculiar medium. Starting from perturbation theory, it is possible to answer this question by means of a Green's function which, being based on local theory, guarantees the propagation of an interaction with the velocity of light.

However, the situation changes if the interaction becomes strong. In this case there arise nonlinear phenomena and strong absorption as a result of inelastic processes. In this connection a possible limitation of the space-time description of the elementary particle structure was indicated several years ago. This limitation is due to the fact that the cross section for an inelastic process does not decrease with increasing energy but tends to a constant limit or even increases slowly. At the same time elastic scattering assumes the character of diffraction scattering on a "black" sphere of dimension a.

When the role of inelastic processes becomes dominant the information obtained is less relevant to the space-time structure than to the production of new particles. The "blackness" of particles makes it impossible to use elastic scattering for studying the space-time distribution of matter. The given example of mesons is a very special case and is therefore not of value in a discussion of geometry.

For the problem we are considering, what would be of interest would be a situation when only "blackness" would appear for the most universal signal. The most universal are weak interactions. Therefore, such a situation can arise if the weak interaction is assumed to increase up to distances dictated by the unitary limit. In this case we expect $a \stackrel{\sim}{=} \Lambda_f$. Thus the possible limitation of local theory by the conditions of the signal propagation inside an elementary particle coincides with the condition following from the existence of a "weak maximon."

8. COMMENTS

The time is not far off when weak interactions in a real experiment will reach the strength of electromagnetic interactions.

The next stage will be the achievement of c.m. energies in the region of 300 GeV by means of colliding beams. If the dominant role of the weak interactions when $E \stackrel{\sim}{=} E_f$ is confirmed (at least for large momentum transfer), then restrictions on coordinate measurability of point-like events, stated in this report, will achieve the importance of a principle.

REFERENCES

1. A. A. Friedman, The World as Space and Time, "Nauka," Moscow (1965).
2. Gravitation and Relativity, ed. by H. J. Chin and W. F. Hoffman, W. A. Benjamin, Inc., N. Y. (1964).
3. J. A. Wheeler, et. al., Gravitation Theory and Gravitational Collapse. University of Chicago (1965).
4. K. Menger, Proceedings of the National Academy of Science, (U.S.A.), 28, 235 (1942).
5. D. I. Blokhintsev, Report JINR P2-6094, Dubna (1971).
6. M. A. Markov, Preprint, JINR E-2644 (1966), E2-5271 (1970), Dubna.
7. J. A. Wheeler, see ref. 3.
8. D. I. Blokhintsev, Uspekhi Physica Nauk XII, 381 (1957).
9. D. I. Blokhintsev, Geometry and Physics of the Microworld. Preprint E2-6652 Dubna (1972).
10. D. I. Blokhintsev, V. S. Barashenkov, V. Grishin, Nuovo Cimento, series X, 9, 249 (1958).
11. D. I. Blokhintsev, Space and Time in the Microworld, (Monograph), "Nauka," Moscow (1970).
12. D. I. Blokhintsev, Nuovo Cimento, Series X, 9, 925 (1958).

INITIAL EXPERIMENTS AT NAL

R.R. Wilson

National Accelerator Laboratory

Batavia, Illinois 60510

For the past several years a number of my stout hearted colleagues have been building a large accelerator close to Chicago near a village called Batavia, Illinois. About a year ago they managed to make it work at its designed energy of 200 GeV. During the past year they have been bringing the energy up, enough so that about a month ago they were able to reach 400 GeV. At the same time they have been working hard to increase the reliability of the machine and to make the intensity larger. About a year ago, when it first came into operation, the intensity was a few times 10^9 protons per pulse, and by now the intensity has increased by several orders of magnitude: it now regularly runs at more than 10^{12} protons per pulse. Thus we are presently ready to embark on the adventure of looking into this new energy region. And adventure it is, because in spite of all of the very respectable work that has been done in this century with the succession of accelerators that started with the Cockroft-Walton machine, in spite of all of the results that have been obtained, the old questions still remain: why the electron has the mass that it has, and why the proton is about 2000 times heavier and has the mass that it has. Why is there a muon? What is the relation between the weak force, the gravitational force, the nuclear force, the electrical force? These are just as much mystery now as they were many years ago; just as much of a challenge to the experimenter as they were when our pioneer forebears started at the beginning of the century. In addition, of course, there are many more problems that are posed by the new symmetry schemes and by the many beautiful results concerning so-called elementary particle states. These results raise new problems for further clarification and hopefully for simplification. Of course one can not expect that with our machine we are going to be able to answer all, or even any, of these problems. Still, we will be gratified if we are to be able to add even some clarity to our understanding of these vital questions.

Now I have a problem in talking about the experimental program at NAL. We have some 200 proposals, of which over 80 have been accepted, and of these work is in progress on about 30. I suspect that the last thing you want to hear is a long list of those experiments. Let me instead rather arbitrarily discuss a sampling of the experimental work we expect to do.

Now I don't intend to talk primarily about the accelerator itself; on the other hand you do have to know something about it in order to understand the flavor of the experiments. To orient you, I'd like to give an impressionistic description of the facility without very much comment until we come to the experimental areas.

The site is about 5 Km wide. Most characteristic

of the project is the main proton accelerator which is
comprised of about 1000 magnets placed to form a ring
which is exactly 1 Km in radius. The accelerator complex consists of several cascading separate machines.
First there is the Cockroft-Walton, then the Linac, then
the Booster Accelerator and finally the Main Accelerator.
Each one, working in its appropriate energy range, feeds
protons to the next, eventually the protons being loaded
into the main ring and accelerated from about 8 GeV when
they come out of the booster accelerator to 200 or 400,
or eventually even to 500 GeV. Once they reach their
final energy, they can either be used within the machine,
or they can be extracted out into an external proton
beam and used in one of three experimental areas which
I will describe later on.

    There are four areas where experiments are made.
The first one is within the machine itself; an internal
target is used. It is felicitous that the first experiment in that area, the first experiment to be made
in the laboratory at all, was a collaboration between
scientists from the USSR and America who studied the
elastic scattering of protons by protons. Apart from
the high energy protons themselves, the most important
part of the experiment is the target, a gas jet target
that was made at Dubna, in the USSR. It consists of a
supersonic stream of molecular hydrogen that is shot
down through the proton beam. The seven Soviet scientists and their wives have been at the Laboratory almost a year now. They installed their target in the
machine this summer, and they and their American colleagues have been working happily with it ever since.
In a sense, perhaps you can say that these are intersecting beams. Protons in their jet at supersonic

speeds make collisions with our high energy protons. Our beam is about a few millimeters in diameter; the Soviet beam, if you will, is about a cm in diameter: at each pulse the energy sweeps from 8 GeV to as much as 400 GeV.

In the experiment, they measure the recoiling proton at an energy of some 100 MeV, which means that they are measuring very soft collisions. This enables them to plot a diffraction curve, and from that they obtain the size of the proton. They have taken data up to 400 GeV. As the energy of the protons is increased, a shrinkage of the peak occurs that indicates that the size of the proton appears to be larger at higher energy than at lower energy. At the very small momentum-transfer points (which they can measure because the hydrogen target is so thin that they can measure very low energy proton recoils) they can get down to the region of Coulomb interaction. Between the regions of Coulomb interaction and nuclear interaction there is a region of interference between the two kinds of scattering. By that interference, they can measure the real part of the scattering as well as the imaginary part which they get from the slope parameter. To get their most recent results, I am afraid you'll have to go to the New York meeting.

There are several other experiments that utilize the same target. In addition, there is a rotating target: a little wheel to which is fastened a thin vane of polyethylene film which rotates through the beam once every revolution. On the same wheel there are whiskers of carbon. The targets rotate through the beam many times as the energy of the protons changes, so it is possible to study the scattering as a function of energy during each pulse. The rotating targets are somewhat thicker than the hydrogen jet target.

Another experiment being done by a group from Wisconsin and Harvard is designed to use the internal target to measure the production of gamma rays, presumably made by π^0 decay. They use a piece of lead glass many radiation lengths in thickness and in front of which is an anti-coincidence detector so that they are sure that gamma rays are measured. In spite of the fact that they are sitting right next to a powerful accelerator in which is circulating a large current of protons, they find that the background is not very high. You would think that this counter would be swamped by neutrons and other particles produced by proton losses in the synchrotron. Not at all. Perhaps it's because the accelerator is spread out over such a great distance. A time-of-flight method is used to verify the signal from the target and to reject the background from the accelerator. The beam in our machine consists of successive clusters of protons, each cluster of protons being about the size of your little finger in thickness and about 30 cm long. The clusters are separated by about 6m. This circumstance lends itself very naturally to timing the signals to an accuracy of 10^{-9} seconds. It turns out that the transformation of the solid angle from the C.M. system to the Laboratory system produces just the right variation so that if the scaling law obtains then the counting rate should not vary with energy. Indeed in the preliminary measurements they have made so far, that's just what they have seen: an approximately constant counting rate of gamma rays, indicating that scaling does obtain.

We also have an NAL group studying the reaction of

protons on protons going to gamma rays plus anything else, James Walker and his colleagues. Their measurements are made in the forward direction, again with a counter telescope not very different from the one just described. They are searching for large momentum transfer events, observing gamma rays at the angle that corresponds to $90°$ in the center of mass system, about 100 milliradians in the laboratory. Measured is the counting rate of the gamma rays as a function of the momentum transfer; they see the counting rate come down with the typical exponential form, but eventually the exponential flattens out. Thus at a momentum transfer of the order of 7 GeV/c, which is a very large momentum transfer, they see a counting rate of gamma rays that is an order of magnitude more than you would expect from the exponential decrease. That's an interesting result in that it appears to agree with the kind of result that Lederman and his colleagues have been observing with the CERN intersecting storage ring. A different kind of measurement entirely is about to be started in the internal target area by Cline and colleagues who will make a search for a curious new particle that would not interact strongly with other particles. Their equipment is necessarily compact and simple. The internal target is particularly applicable to such a particle search because the energy of the protons sweeps through its full range during each acceleration cycle, hence a new particle should exhibit an identifiable threshold of production.

Let me turn now to the external proton beam that can be led to three different experimental areas. The protons can be extracted from the accelerator in a rapid mode by "pinging," i.e., suddenly deflecting the internal beam so that it enters a deflecting field that guides it tangentially out of the accelerator. In that case the beam can have a duration of from a few to about twenty microseconds.

If a beam of larger duration is desired, then the guide field of the accelerator is adjusted so that the radial oscillations of the proton orbits (betatron oscillations) are caused to grow exponentially. This is done by making the number of betatron oscillations per turn a half integer. Eventually the orbit oscillations grow large enough that the protons pass into an extraction channel and are then deflected out of the accelerator. This is a continuous process so that the external beam obtained by this method can have a duration of several hundred milliseconds.

After extraction the beam is led through a maze of tunnels, the "switchyard," the purpose of which is to direct the beam to one of the three separate experimental areas. Thus the protons can be allowed to proceed straight ahead to the Neutrino Area, or they can be switched to the left (west) to the Meson Area, or to the right to the Proton Area - the names being arbitrarily chosen, yet somewhat characteristic of the principal kinds of measurements to be made in a given area.

Let us first consider the Neutrino Area. The protons are sent to the surface and strike a target, usually a narrow piece of steel, placed within a large mound of earth shielding. Many particles are made in the target in the showers that result from proton collisions with the nucleus, but let us consider the pi mesons that result. These can be directed by means of magnets to a long, evacuated decay pipe - about 1000 feet long. As the pions go down the pipe, they decay into muons and neutrinos. The muons as well as all the other kinds of strongly interacting particles strike the end of the pipe and are absorbed in a mound of earth that is almost one kilometer in length. The neutrinos of course can easily penetrate the mound so a pure neutrino beam emerges from the far end of the mound.

If a raw beam of pions, i.e., one containing pions of all energies is incident on the decay pipe, then neutrinos of all energies emerge. However, if the pions from the target are sorted magnetically so that pions of a narrow band of energies go down the pipe, then the resulting neutrinos are also essentially monoenergetic. Actually because kaons are also present and also decay by a neutrino mode but of different energy, there are two bands of neutrino energy. These neutrino beams, with energies up to several hundred GeV, are one of our principle research tools.

From work on the interaction of neutrinos done at other laboratories, it is known that the cross section for the interaction increases almost linearly with energy up to about 5 GeV. There are also data up to about 10 GeV that are consistent with this linear rise, but the statistical accuracy is not very good as yet. Our experimenters are recording hundreds of events in the hundred GeV range and I am anxiously awaiting their first results to see whether the linear rise continues or whether the curve bends over.

One reason for it bending over would be the existence of the intermediate boson or W particle. Such a particle, if it is made copiously enough, would reveal itself in the experimental equipment by showing two muons being produced at once, one positive and one negative. That would indeed be exciting, but there are numerous, if more mundane, kinds of measurements that will also reveal a vast amount about the weak interaction. A Cal-Tech group is making the bi-chromatic measurement while a Harvard, Pennsylvania, Wisconsin collaboration will use a broad band of neutrinos. Both are ready to start measurements.

At the same time that neutrinos are made, of course muons are also made. These are led magnetically from the decay pipe to a beam line that goes along the side of the berm which filters the neutrinos. The beam line ends in what we call the Muon Laboratory, and here an experiment is underway by a Cornell-Michigan group to test the scaling laws of inelastic muon interactions in the new hundred GeV energy region. First results should be forthcoming before too many months. The Chicago Cyclotron has also been moved to this area and will be used as a giant spectrometer for detailed studies of the muon interactions.

Two hydrogen bubble chambers are located at the end of the Neutrino Area. A 15 foot bubble chamber is nearing the end of construction and is placed directly in the neutrino beam. A 30 inch hydrogen chamber was moved from Argonne National Laboratory to a position next to the new 15 foot chamber. It was in operation a little before a beam was obtained and has been used since the first external beam was obtained last summer. Protons or other particles from the neutrino target can be led to it by a beam line that goes along the other side of the neutrino filter from the muon beam. Preliminary runs have been made using protons at 100, 200, and 300 GeV and using negative pions at 200 GeV. Much of the results of these bombardments has already been published. Because the particles of the backward hemisphere of a collision appear at relatively low energy in the laboratory system, most of the interaction can be studied very effectively using the 30" bubble chamber. A short pulse of protons is made at the end of the slow pulse and these protons can be by-passed to the 30" bubble chamber. Thus, in principle, we should be able to get a picture parasitically for each pulse of the accelerator. Indeed, most of our running of the bubble chamber is in this mode, although the inefficiency of doing many things at once sometimes gets to be pretty frustrating for the experimenters. They have also been able electronically to identify the kind and position of particles entering and leaving the bubble chamber. When fully developed, this will constitute a powerful tool for the study of hadronic interactions.

Now let me turn briefly to the Meson Area which is just being completed and which will be brought into operation in the next months. The idea for this area is to have the protons strike the target at 300 GeV, then to have about six separate beam lines so that the secondary particles, usually mesons, can be studied. The experiments in this area have more nearly the character of experiments done at lower energies in the other laboratories. Thus a quark search will be made here. In the neutral beam, the neutron-proton interaction will be studied and later a K_o regeneration experiment will be made. One of the charged particle beam lines will eventually become part of an elegant spectrometer that can be used for precise studies of meson interactions at high energy.

On the other side of the neutrino beam is the Proton Experimental Area. The proton beam instead of being brought up to the surface, as in the other two areas just discussed, is raised from a depth of about 20 feet underground characteristic of the depth of the accelerator to a depth of about 10 feet. The idea is that the shielding of the earth itself can then be used rather than piling up concrete blocks ad infinitum. Thus a series of pits have been dug into which the full intensity of the proton beam can be directed onto various kinds of targets. The area is also designed to receive protons of the highest energy that we can make, perhaps even 500 GeV, as compared to the Meson Area which was designed for use with 200 GeV protons but which has now been hardened to a maximum of 300 GeV.

It is anticipated that the scattering of protons by protons will be studied at high momentum transfer in this area. Also being mounted is an experiment to search for electrons that might be emitted at large angles. In another experiment an exploratory search will be made to observe anything coming out at large momentum transfer - both of these experiments might explore the depths of the proton. In this same area a raw photon beam will be made in order to explore photoproduction in the hundred GeV range and, later a precision electron beam will be produced in order to carry electron physics into that range.

In another year or two I would hope that the measurements of the scientists using our facility will have progressed far enough that we can begin to answer some of the old questions about the structure of elementary particles and about the forces that act between them.

Behram Kursunoglu presented his paper "Fundamental Interactions".

TOTAL ABSORPTION DETECTORS AND THEIR APPLICATIONS TO HIGH ENERGY PHYSICS

Robert Hofstadter

Department of Physics and High Energy Physics Lab., Stanford Univ., Stanford, Calif. 94305

INTRODUCTION

My purpose in this paper is to suggest that it is possible to bring precision to the processes of detection of particles and photons of very high energies. I have felt for a long time that high energy physics will someday turn into a spectroscopic subject and at that time high resolution methods will be essential to further developments. In the past few years I have been trying to advance the detection technique in two ways. One involves the straightforward development of a precise, large-solid-angle magnetic spectrometer capable of resolving momenta to one part in ten thousand. The second technique is concerned with the total absorption method of detection, and this is the actual subject of the present paper.

One may wonder why precision and high resolution methods are needed in high energy physics. As the very simplest example, I wish to point out that to resolve an elastic peak, e.g., in proton-proton scattering at

500 BeV, from an inelastic peak involving the production of a single pion, one needs a resolution of approximately $0.137/500 = 2.7 \times 10^{-4}$. One can obtain such a figure at high momenta with a small-solid-angle magnetic spectrometer but probably no other technique is today capable of such high resolution. On the other hand, in the backward region of scattering, near $180°$ in the CM system, a lower momentum magnetic spectrometer is quite usable, and that is one reason for the development of the spectrometer mentioned in the first paragraph.

I think it is a highly desirable goal for the future of detection technology to meet the needs I have outlined. Our own efforts are pointing toward this direction. At the present time a resolution figure of about $\pm 35 \times 10^{-4}$ has been achieved at the High Energy Physics Laboratory at Stanford in detecting gamma rays, electrons and positrons of energy 15 BeV. In the detection of hadrons, the corresponding limit we have achieved is a meager $\pm 1200 \times 10^{-4}$ for 14 BeV protons or pions. Therefore there is much room for further improvement, particularly in the case of hadron detection.

These results have been achieved by using the method of "total absorption" of a particle's or photon's energy. We have developed two acronyms for such detectors: TASC and TANC detectors. TASC refers to a "total absorption shower counter" and TANC to a "total absorption nuclear cascade" detector.

Symbolically Figures 1 and 2 show the main ideas in the total absorption process of detection. The general idea is to break up the original energy into a large number of small parts each of which is readily absorbed. In TASC detectors the process is rather simple because of the "shower" mechanism. The distance for splitting

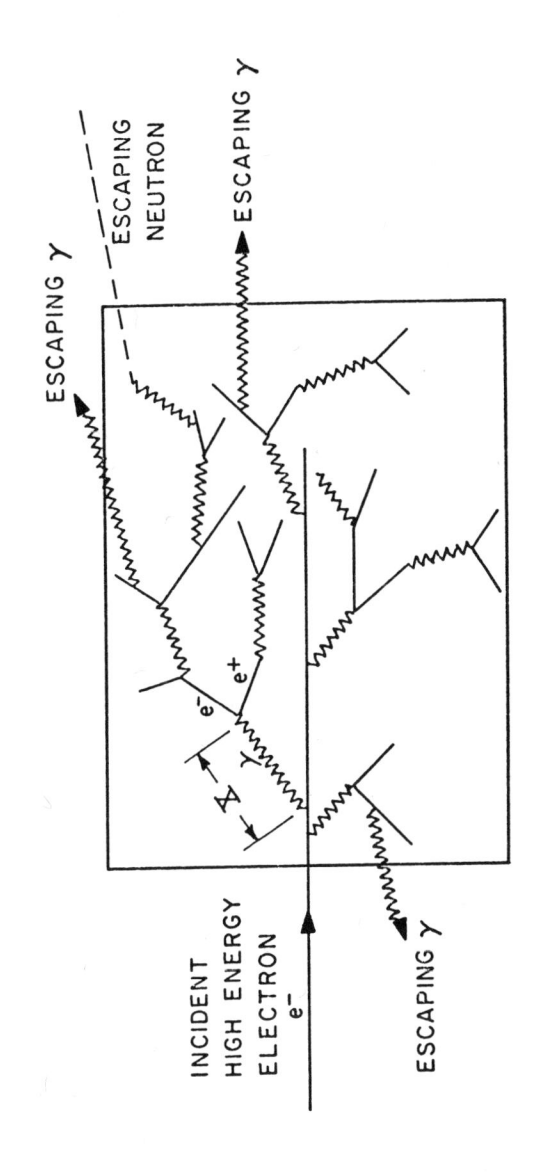

Figure 1. The mode of operation of a TASC detector system is shown in this figure. Escaping gamma radiation is of little consequence if the crystal is of large volume. In NaI(Tl) the distance X is on the order of one inch.

Figure 2. This figure shows a schematic view of the detection process in a TANC detector for hadrons. Note that some neutrons and pions may escape from the detector if their mean free paths exceed the distances to the sides of the crystal. It is also true that some binding energy lost in nuclear collisions may not be recovered during the scintillation pulse time.

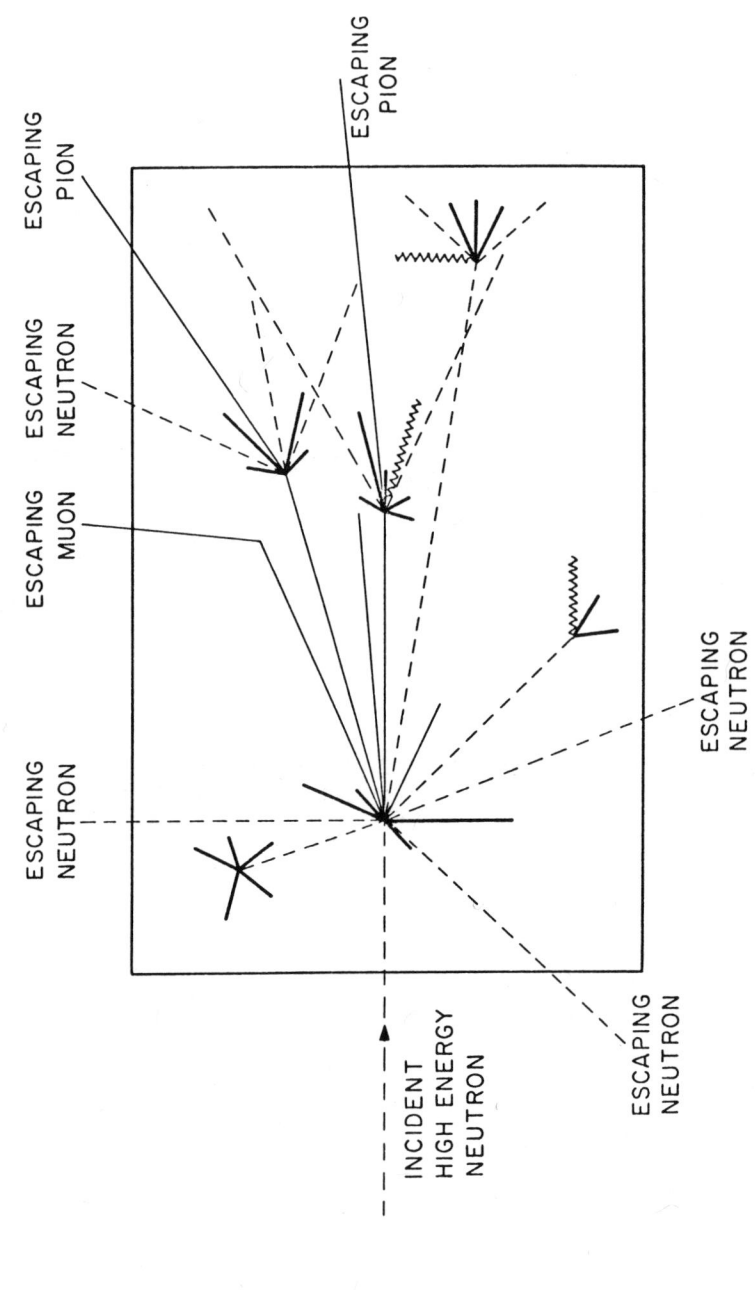

Figure 3. This figure shows a composite 16" diameter × 24" long crystal made of three crystal blocks. The mounting house is shown below.

the energy into approximately a factor of two is about one inch in the scintillator NaI(Tl). After twenty such divisions all the energy is easily absorbed. Figure 3 shows a NaI(Tl) crystal and its housing that has performed very well as a TASC detector. In Figure 4 we see the response of this scintillation counter to energetic electrons of energy 15 GeV. This result represents the best resolution we have obtained so far for "electromagnetic" particles. Gamma rays will behave in the TASC detector in almost exactly the same way as electrons. At lower energies the resolution is not as good. For example, at one GeV the FWHM is about 3%. At energies higher than 15 GeV the resolution should improve, probably as the inverse 0.25 power of the energy. Thus at 500 GeV one should be able to achieve a resolution figure of about $\pm\ 12 \times 10^{-4}$. If we learn more about the light collection in large crystals this figure can probably be improved.

In the case of TANC detectors for hadrons, escaping neutrons, muons and neutrinos will probably furnish the main limit to the resolution. However at high energies (>1000 GeV) theory indicates that it is likely that 85% or more of the cascade turns into electromagnetic forms of energy (showers) and these are easily absorbed. Thus the resolution at very high energies may be better than might be estimated at first sight.

Using a large TANC assembly of NaI(Tl) crystals as shown in Figure 5 a symmetrical absorption peak for 14 GeV/c pions has been obtained of full width half maximum 24% (or less). Approximately 85% of the total energy is absorbed in the assembly. The TANC detector contains about 3.5 tons of NaI(Tl). At energies of a few hundred GeV an anticipated FWHM of about 5% may be

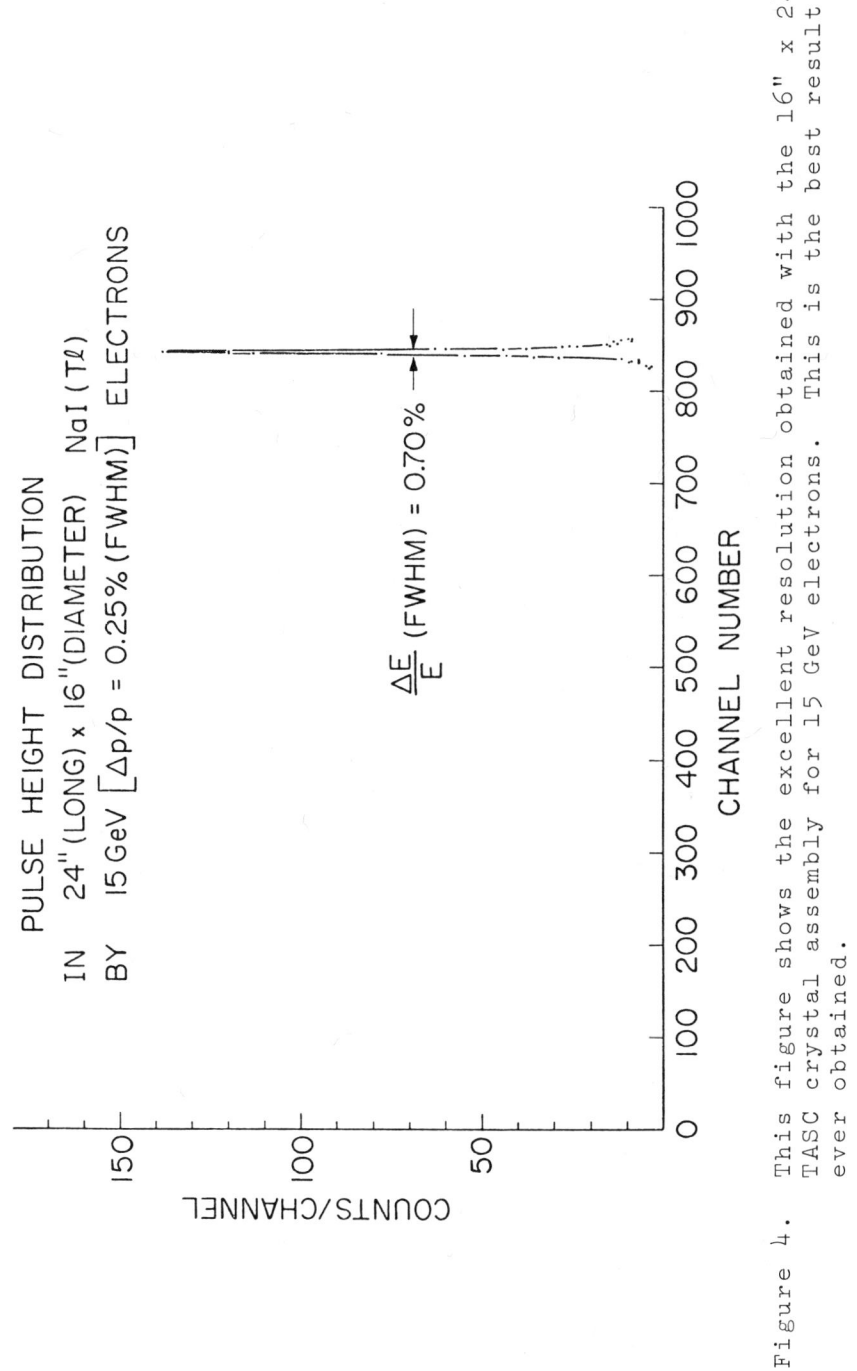

TOTAL ABSORPTION DETECTORS

PULSE HEIGHT DISTRIBUTION
IN 24"(LONG) x 16"(DIAMETER) NaI(Tℓ)
BY 15 GeV [Δp/p = 0.25% (FWHM)] ELECTRONS

$\frac{\Delta E}{E}$ (FWHM) = 0.70%

Figure 4. This figure shows the excellent resolution obtained with the 16" x 24" TASC crystal assembly for 15 GeV electrons. This is the best result ever obtained.

Figure 5. This large assembly of a TANC system was used to provide the data which give a symmetrical total absorption peak with a FWHM of 24% or less and an absorption figure of 85% of the total energy.

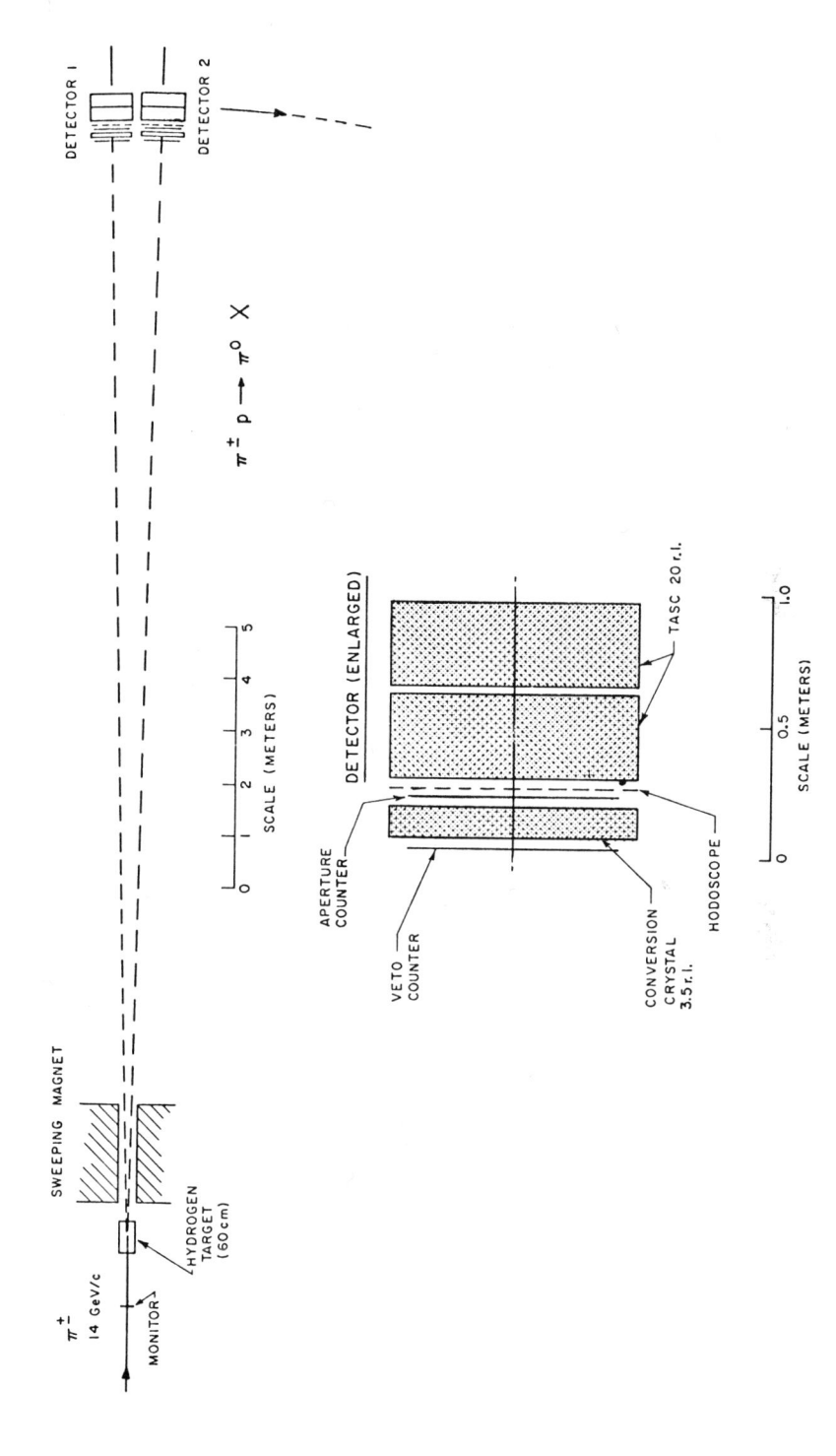

TOTAL ABSORPTION DETECTORS 117

Figure 6. The experimental arrangement for the inclusive π^0 experiment is shown in this figure.

obtained.

I would now like to give a short description of some experiments in high energy physics to which these techniques can be applied.

An experiment was recently carried out at the Stanford Linear Accelerator Center (SLAC) using the experimental arrangement shown in Figure 6. In this experiment, the incident pions had momentum 14 BeV/c, and it was desired to make measurements of the inclusive reaction:

$$\pi^{\pm} + p \rightarrow \pi^{0} + \text{anything} \tag{1}$$

$$\pi^{0} \rightarrow 2\gamma$$

The π^0 mesons were detected by their practically immediate transformation into two energetic gamma rays. The dashed line in Figure 6 near detector 2 shows that the detecting system can be moved as a function of angle about the direction of $0°$ (forward angle). Each detector assembly is of the type shown in the lower part of Figure 6. The veto counter assures us that only gamma rays are detected, and the hodoscopes measure the angles at which the neutral pion's γ-rays are received. The equivalent mass of the π^0 is obtained from the equation

$$M_o c^2 = \sqrt{2E_1 E_2 \sin^2 \theta/2} \tag{2}$$

where E_1 and E_2 are the energies of the two simultaneously observed γ-rays and θ is the included angle. The accuracy in determining M_o in this experiment is determined by the hodoscope "grain". Thus the observed width of the π^0 mass is as shown in Figure 7, and depends

TOTAL ABSORPTION DETECTORS 119

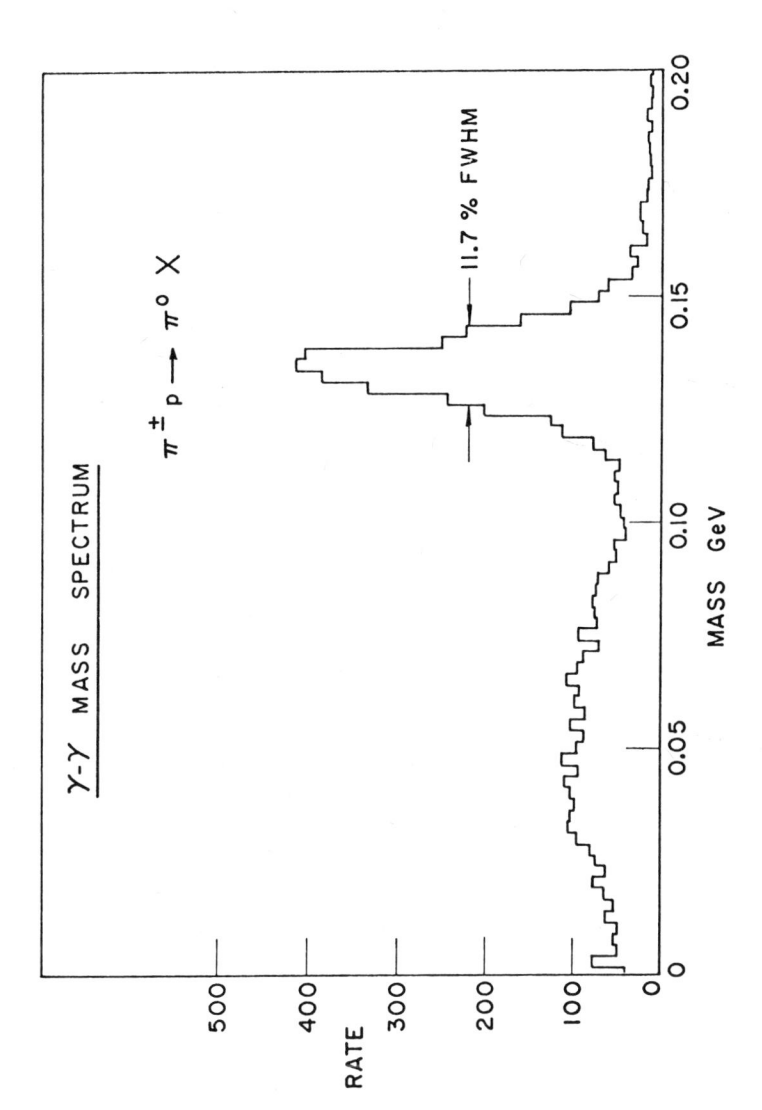

Figure 7. The mass spectrum of the two gamma ray events in the inclusive π^0 experiment. The mass of the π^0 can be identified in the peak where the width is 11.7%.

mostly on the resolution of angle by the hodoscope and is <u>not</u> limited by the energy resolution of the NaI(Tl) detectors.

Observation of π^0 peaks such as they are in Figure 7 allows energy measurements ($E = E_1 + E_2$) to be made and data for an energy distribution in the forward direction can be assembled. Such a preliminary accumulation of data is shown in Figure 8 for incident π^+ and π^- pions. A detection efficiency curve is also shown in the figure and is not yet folded into the final data. It is clear however that we have obtained experimental data which can be compared with modern theories for inclusive reactions such as those proposed by Feynman and by Yang.

A new application of the TASC detector is now being developed and apparatus is now being assembled for use with the colliding electron-positron facility (SPEAR) at SLAC. Our idea is to measure simultaneously the cross section for the four following reactions:

$$e^+ + e^- \rightarrow e^+ + e^- \qquad (3)$$

$$e^+ + e^- \rightarrow 2\gamma \qquad (4)$$

$$e^+ + e^- \rightarrow e^+ + e^- + \gamma \qquad (5)$$

$$e^+ + e^- \rightarrow \mu^+ + \mu^- \qquad (6)$$

Such an experiment will test the validity of quantum electrodynamics at a center of mass energy starting at about 2.0 BeV and going up subsequently to 4.5 BeV. If the experiment is successful QED will be tested at distances as small as 2×10^{-15} cm. The proposed experimental arrangement is shown in Figure 9. It is to be

TOTAL ABSORPTION DETECTORS

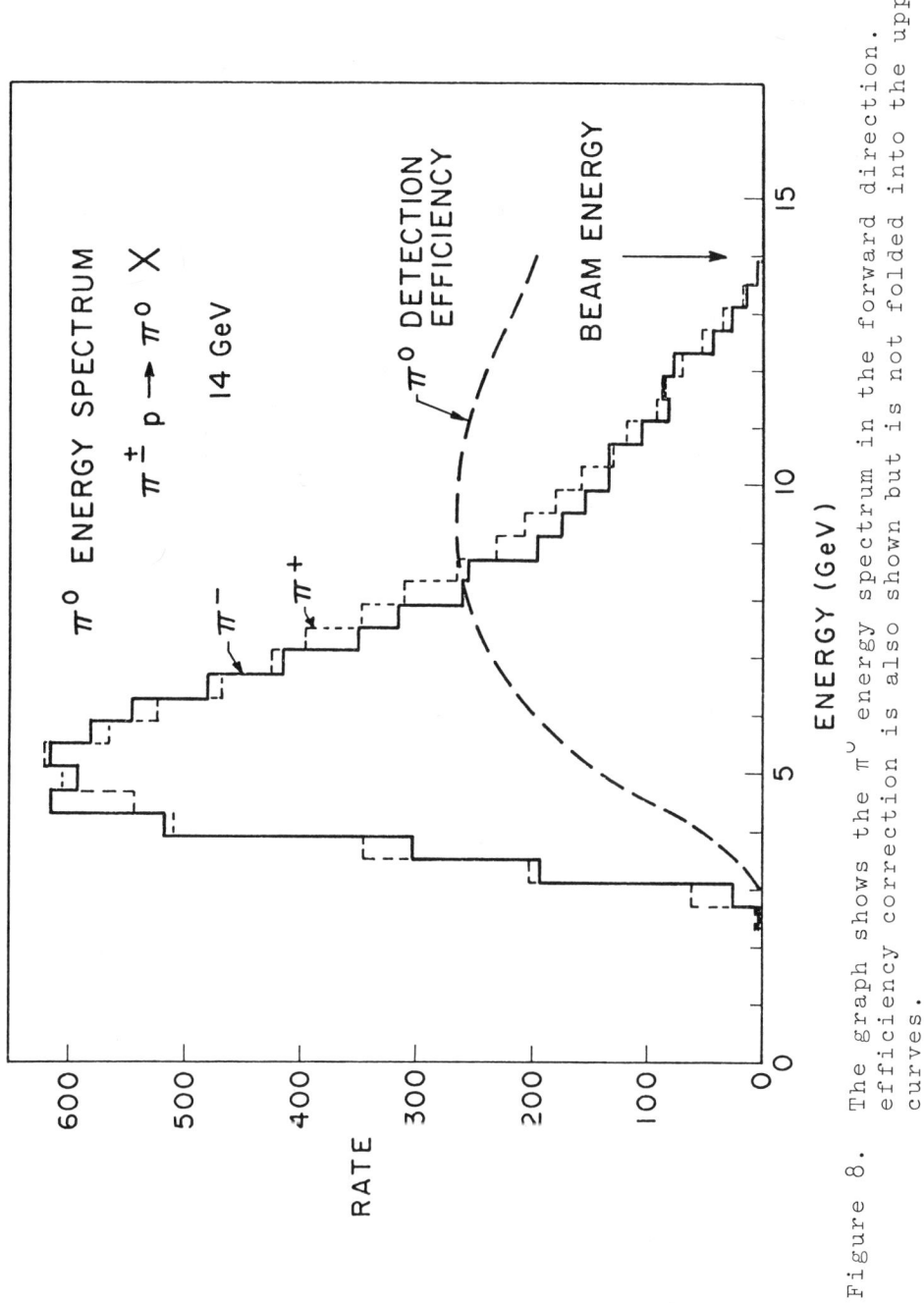

Figure 8. The graph shows the π^0 energy spectrum in the forward direction. An efficiency correction is also shown but is not folded into the upper curves.

SPEAR DETECTOR

$e^+ e^- \rightarrow \begin{cases} \gamma\gamma \\ e^+ e^- \\ e^+ e^- \gamma \\ \mu^+ \mu^- \end{cases}$

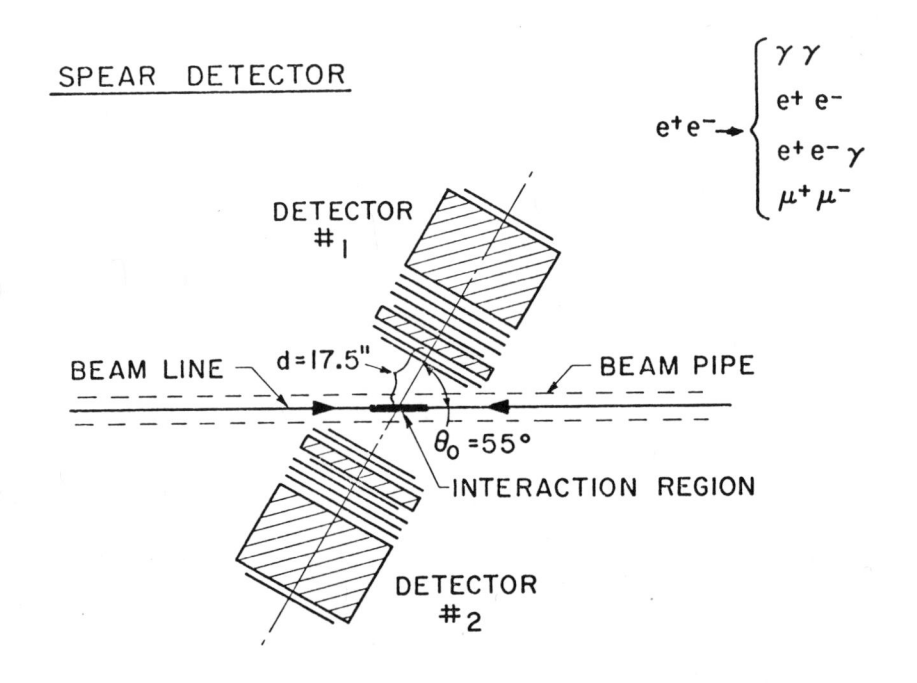

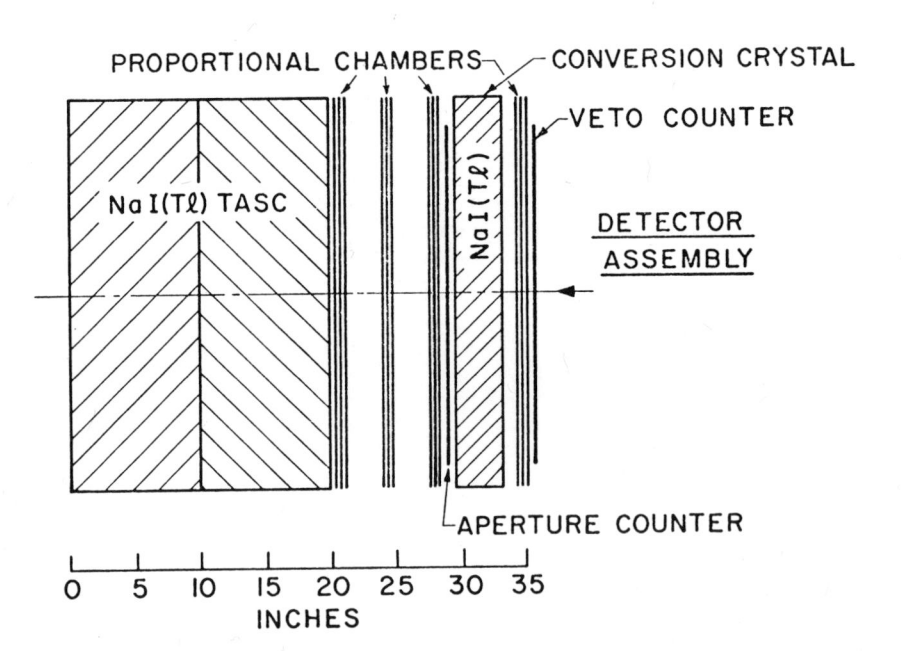

Figure 9. This figure shows the design and arrangement of the colliding beam experiment at SPEAR.

noted that a comparison of how well muons and electrons agree in behavior can come out of this experiment. Our experimental group hopes to complete this experiment in 1973.

ACKNOWLEDGMENT

I wish to thank my colleagues in the TASC and TANC research groups at the High Energy Physics Laboratory at Stanford for their kindness in allowing publication in this article of some of the newly measured results. My colleagues include, B. L. Beron, J. F. Crawford, R. L. Ford, E. B. Hughes, L. H. O'Neill, R. F. Schilling and R. Wedemeyer.

Much of this work was supported by the National Science Foundation Grant No. 22880 and NASA Contract NGR-05-020-452. I also wish to gratefully acknowledge the assistance provided by a grant from the General Motors Research Laboratories.

Robert Hofstadter spoke at the Conference dinner hosted by Mr. and Mrs. R. Kirk Landon.

OPENING COMMENTS ON GAUGE THEORIES

        B.W. Lee

        Department of Physics

        State University of New York at Stony Brook

        Stony Brook, Long Island, New York 11790

    I wish to take the prerogative of a discussion leader to make a few opening comments on the subject of gauge theories of weak and electromagnetic interactions to set the stage for this session.

    At the NAL conference last September, we assessed the progress in this field that had taken place in the preceeding year. I am happy to have been able to secure the participation of today's speakers who have led the developments in this field since then. I am especially delighted to have the presence of Professor Steven Weinberg and Professor Abdus Salam, who started this marvelous enterprise.

    Last September, I mentioned the work of Bars, Halpern and Yoshimura as a possible new avenue of model building, incorporating both strong, weak and electromagnetic interactions. I must note here that a parallel development took place at Utrecht, in the hands of Mr. Bernard de Wit. The models of this genre treat the hadronic world and leptonic world as separate up to a point, each having its own set of gauge bosons: the two worlds communicate to one another through the intermediary of a new kind of Higgs mesons which carry both leptonic and hadronic quantum numbers and whose vacuum expectation values are responsible for the coupling of the two kinds of gauge bosons, in much the same way as envisaged by T.D. Lee, Weinberg and Zumino some time ago. There are a large number of questions one can raise about the models of this type: for example, is there a lepton-hadron universality in parity violating weak decays? How does the Cabibbo angle arise in such a model (actually this is the question De Wit addressed himself)? What are the phenomenological manifestations of the new kind of Higgs mesons? Undoubtedly Professor Martin Halpern will discuss these and other related questions.

I would characterize the recent advances in this field as the sharpening of desiderata in model building. Our experiences have shown that, with the limited amount of experimental information available to us now and in the near future, the possibilities in model building are myriad. Secondly, the _classical_ models based on either U(2) or SU(2) are somewhat lacking in esthetics when hadrons are incorporated in them. For these reasons, instead of waiting for elucidation from our experimental colleagues, a number of theoreticians have discussed and are pursuing various theoretical propositions that might be true and desirable in the ultimate model. These include the possibilities that the electron mass is computable; that higher order effects are responsible for spontaneously broken gauge symmetry; that the Higgs phenomenon occurs due to a collective motion of fermion fields without the necessity of introducing a set of scalar fields; that approximate broken hadronic symmetries, such as the broken chiral symmetry and PCAC and the lore of current algebra, have a deeper rationale in a gauge theory of weak and electromagnetic interactions. One reason for this line of investigation is the hope that somehow these considerations will lead to a unique and esthetic model of elementary particles. The Cambridge trio of Professors Sheldon Glashow, Sidney Coleman, and Steven Weinberg, who spearheaded this line of inquiry, will discuss their respective views and works.

Lastly, I would like to say a few words about the question of renormalizability of spontaneously broken gauge theories. The systematics of renormalization in terms of Ward-Takahashi identities have been worked out for a wide class of gauges. These discussions presume the existence of a gauge-invariant regularization. A very elegant regularization method was offered by 't Hooft and Veltman, which is based on the idea of analytically continuing Feynman integrals in the number of space-time dimensions. Personally, I think that this method regularizes <u>all</u> Feynman diagrams in a gauge-invariant way. However, a rigorous proof of this fact, in a manner that is acceptable to a mathematician, as the proofs of Hepp, and Speer were for other methods of regularization, is still missing and clearly desirable.

Steven Weinberg was awarded the J. Robert Oppenheimer Memorial Prize in an evening ceremony.

UNIFIED GAUGE THEORIES OF STRONG, WEAK AND ELECTROMAGNETIC INTERACTIONS

M. B. Halpern

University of California

Berkeley, California

I. INTRODUCTION

I am going to try to sketch the form of a recent unified gauge (renormalizable) theory of strong, weak and electromagnetic interactions,[1,2,3] and report briefly on some new developments. Before giving any details, I would like to go over a few of the more salient features of these models.

1. <u>Data Fitting</u>. To my knowledge, the models appear to fit all known low energy data (say less than one GeV) for all strong, weak and electromagnetic interactions.

2. <u>Natural and Fairly Realistic Strong Interactions</u>. There are a number of different models[4] proposed by other workers which can make a claim similar to (1) above. The distinction of our model is that the strong interactions are treated naturally, and somewhat realistically. For example, the usual $(3,\bar{3})$ broken $SU(3) \otimes SU(3)$ strong interaction is taken as a starting point in our approach. Thus we have the usual octets of vector and axial vector

mesons, pions, etc., as well as the usual PCAC, and so on. In fact, I believe the hadron theory by itself realizes the original ideas of Yang and Mills;[5] more concretely, we take over bodily all "good" results of the old algebra-of-fields-effective-Lagrangian approach, now in the context of renormalizability.

3. <u>Lepton Theory</u>. By requiring the "known" strong interactions, only certain lepton theories are allowed. Weinberg's theory is the most natural, but one other model, midway between Weinberg and Georgi-Glashow is allowed.

4. <u>The M-particles</u>. Perhaps the most salient feature of these models is the required existence of some new scalar and pseudo-scalar particles (a "nonet" of each). Originally derived[1] from the requirement of renormalizable hadronic vector meson physics, these particles have later played a larger role in the overlap region of all three fundamental forces.[2] Before spontaneous breakdown, e.g., these particles are the <u>only</u> ones coupling with all three fundamental interactions. After spontaneous breakdown, they are responsible for (a) hadronic vector meson masses, (b) known semi-leptonic processes and (c) known hadronic symmetry breaking. The masses of the M's can be taken very high if necessary, but it would be much more exciting if they were found at relatively low mass. We will return to this subject later.

5. <u>Regge and Dual Models</u>. Of course, our hadrons do not lie on linear trajectories, and so on; so our hadron physics is by no means totally realistic. On the other hand, work in dual models[7] has shown that <u>models of our type are precisely the flat-slope limit of dual models</u>. I like to think that our hadrons are the "bottom" of the future dual model - and that workers in gauge and dual models have a great deal to learn from each other.

UNIFIED GAUGE THEORIES

Along these lines I am investigating the possibility that the gauge hadron model Reggeizes in the sense of Gell-Mann et.al[8]. I would interpret the gauge-theory loops as recovering part of the non-zero slope of the dual models. In any case, I would guess that these gauge Lagrangians are about as good as you can do for hadrons within the ordinary Lagrangian framework.

6. <u>Scaling</u>. Although the currents of the model resemble algebra of fields (no fermion "constituents"), the model scales in perturbation theory[9] as well as does the gluon model. This is a surprising high-energy bonus, and may lead to a reinterpretation of electroproduction data. I will also say more about this later.

II. SOME DETAILS OF THE HADRON MODEL

I think the historical viewpoint is very instructive in presenting the model. About a year and a half ago, Bardakci and I set ourselves the task of constructing a gauge model of the known low spin hadrons,[1] following the ideas of Higgs-Kibble[10] and Weinberg-Salam.[6] I cannot <u>prove</u> uniqueness of this scheme, but I can say that, to this date, no one has found any alternatives to this type of scheme. <u>The crucial ingredients are a new set of pseudo-hadron (scalar and pseudoscalar) fields</u> called $M_{L,R}$ which you can think of as 3×3 (or later 3×4) matrices. Under a set of locally realized $SU(3) \otimes SU(3)$ transformations, these fields transform as reducible fundamental representations:

$$M_{L \atop R} \to S_{L \atop R}(x) M_{L \atop R}, \quad S_{L \atop R}(x) = \exp\{i\alpha_{L \atop R}(x) \cdot \lambda\}$$

where the λ are the usual SU(3) λ-matrices. Thus the (3 x 3) matrices are transforming as three right and three left handed triplets under the local group. They couple to the vector and axial vector mesons via the usual covariant derivatives:

$$\Delta_\mu M_{L \atop R} = \partial_\mu M_{L \atop R} - if V_{L \atop R} M_{L \atop R}$$

where f is the universal vector meson coupling.

If, under spontaneous breakdown, the M's pick up a vacuum expectation value $<M>$ k ⊗ 1, it is easy to see that all 16 hadronic vector mesons are raised to the common mass $f^2 k^2$. If, in fact, the Lagrangian is assumed invariant under the additional <u>global</u> SU(3) ⊗ SU(3)

$$M_{L \atop R} \to M_{L \atop R} (S')_{L \atop R}^{-1}$$

then the resulting theory is completely SU(3) ⊗ SU(3) symmetric after spontaneous breakdown. The <u>final</u> symmetry is the product group $F_\alpha \oplus F'_\alpha$, where the F's are the generators of the unprimed (local) and primed (global) groups respectively.

Alternatively, the final symmetry may be broken via "insertions" in the unprimed group, along with more general $<M>$. Pions ($\Sigma \equiv \sigma + i\pi$, $\Sigma \to S_L \Sigma S_R^{-1}$) may be added in much the usual way to generate Goldstone-broken symmetries, and so on.

Except for the extra M's then, the theory is satisfactory, and almost ordinary. We confine ourselves here to a few remarks on the M's. (1) In the absence of a different model, the existence of the (reducible) M's is a direct consequence of known hadron structure: The degree of reducibility (3 or more columns) is fixed by requiring the usual massive hadronic vector mesons.

(2) The M-system has a built-in "open-door" for the non-strong interactions to enter. The hanging indices (acted upon by the transformations F'_α) are at our disposal. The obvious suggestion is to realize non-strong interactions as a local sub-group of the F'_α.

(3) Under the final physical $SU(3) \otimes SU(3)$ ($F\alpha + F'_\alpha$), the M's no longer transform as triplets. Under these two-sided final symmetry transformations $M \to SMS^{-1}$, so the remaining M's form octets and singlets of scalars and pseudoscalars.

III. INTRODUCTION OF NON-STRONG INTERACTIONS

Our path is then to start with the "known" strong interactions, and see what we can learn about the non-strong, taking them as a local subgroup of the primed group; thus (Ur-)hadrons $\to M_L^R \leftarrow$ (Ur-)photon, W's. "Ur" is appropriate here, because in fact, final physical vector mesons are linear combinations of these. (The mixing is small).

The M-system places some important restrictions on the Ur-"weak" vector mesons. For example, by considering the <u>charges</u> of the M's, we can deduce that there must be <u>at least four</u> of these non-strong vector mesons coupling from the primed side.

We know at least that the diagonal entries of the M's are neutral because we need their vacuum expectation values for the hadrons. Let's call Q_H and Q' the Ur-charge operators for the strong and non-strong systems. We know two things then. Q_H is the usual quark form

$$Q_H = \begin{pmatrix} 2/3 & & \\ & -1/3 & \\ & & -1/3 \end{pmatrix}$$

while we must require

$$Q_H \langle M_L \rangle_R - \langle M_L \rangle_R Q' = 0$$

This last is the statement that our required vacuum expectation values are neutral. Thus,

$$Q' = \begin{pmatrix} 2/3 & & & \\ & -1/3 & & \\ & & -1/3 & \\ & & & \ddots \end{pmatrix}$$

where the dots (not yet determined) are useful when we take the M's as 3 x 4. This Q' is the charge to which the Ur-photon couples. It is simple to convince yourself, in the same way you would for hadrons, that this Q' cannot be a pure SU(2) generator. Hence at least an SU(2) ⊗ U(1) of non-strong vector mesons is required.

Once this is understood, we encounter the problem of suppressing the neutral $\Delta S = 1$ currents in the presence of the Cabibbo angle. This we were not able to do while embedding the local SU(2) ⊗ U(1) in a primed U'(3) ⊗ U'(3). The next simplest choice for the primed group is U'(4) ⊗ U'(4). The subgroup that we choose to realize locally corresponds to the 4 x 4 matrices

$$\tau^\alpha = \tfrac{1}{2}\begin{pmatrix} \hat{\tau}^\alpha & 0 \\ 0 & \hat{\tau}^\alpha \end{pmatrix} \qquad\qquad \hat{\tau}^\alpha \equiv \tau_2 \tau^\alpha \tau_2$$

where τ^α ($\alpha = 0,1,2,3$) are the ordinary Pauli matrices. In particular, the <u>left</u> handed triplet of W mesons couples to $(F'_L)_{1,2,3}$, ($\alpha = 1,2,3$), while there is a B^μ, coupling to

UNIFIED GAUGE THEORIES

$$Y' = F'_{3R} + \tfrac{1}{2}(F'_{OL} + F'_{OR}).$$

Notice that, defining

$$Q' \equiv F'_{3L} + Y' \sim \begin{pmatrix} 2/3 & & & \\ & -1/3 & & \\ & & -1/3 & \\ & & & 2/3 \end{pmatrix}$$

gives back our requisite primed charge operator. Cabibbo rotation is defined in the usual way as the usual rotation in the 2, 3 subspace of these matrices. The point is that we have chosen our neutral operators proportional to unity in this subspace, e.g.

$$t_3 = \tfrac{1}{2}\begin{pmatrix} +1 & & & \\ & -1 & & \\ & & -1 & \\ & & & +1 \end{pmatrix}$$

so that, in fact, neutral vector mesons do not rotate at all under Cabibbo rotation, while charged operators rotate as usual. Hence, there are no neutral $\Delta S = 1$ currents in lowest order. These currents are also suppressed in higher order. The argument can be made in strict analogy to such a suppression in the 4-quark theory of Weinberg et al[4] - our four columns of M play exactly the same role.

Further Structure. In this notation, the W's and B can be taken to be Weinberg's and to couple to his leptons in just the usual way. The notation remains pretty, but technical, so I will not go into much of it here. A number of remarks are in order however. (1) Because V's couple to the left of the M's and W's to the right, the Lagrangian contains terms of the form:

$$\mathcal{L}_I = \mathrm{Tr}(VM\tilde{W}M^+)fg$$

where the "twiddle" operation is the usual Cabibbo rotation. Thus when $M \to M + K$, terms of the form $V\tilde{W}$ are induced. These are the usual vector-meson dominated weak interactions.

(2) The "real"-universal-zero-mass photon is a linear combination of the Ur-$(\rho\omega\phi + B,W_3)$ system. An amusing fact learned here is that <u>the strong interactions are stronger than electromagnetic</u> ($f^2 > \frac{4}{3}e^2$). This occurs just as in Weinberg's theory; one shows during the "weak" vector meson diagonalization that $g > e$. Here we have an extra diagonalization to do, but in both cases the asymmetry is a real one because the photon is distinguished as having zero mass. Such a prediction is unique to our type of gauge model.

(3) Thus far, we followed a line of "inducing" non-strong from strong interactions. Now for the first time, we get a "leptonic" feedback to the strong interactions: In our notation, Weinberg's scalar field ϕ transforms as $\phi \to S_L' \phi S_R'^{-1}$. This allows the construction of a $(3,\bar{3})$ hadron symmetry breaking term $[\text{tr}(G\phi^+ M_L^+ \Sigma M_R) + \text{h.c.}]$ Here G is a numerical diagonal insertion matrix. Now, spontaneous breakdown in the ϕ, M system generates a linear term in σ. In these models, the M's must be considered hadrons, but even so, we see that hadron symmetry breaking proceeds with dimension $D \leq 3$. This is a consequence here of known properties of the weak interactions.

(4) <u>Other Lepton Models</u>. We established that four non-strong vector bosons must couple to the M's, but not necessarily to the leptons. We can use this fact to construct an alternative lepton model: couple only the SU(2) to the leptons of Georgi and Glashow. (B couples now only to the M's). Georgi and Glashow's scalar fields are now needed in addition, but primarily to give their

UNIFIED GAUGE THEORIES

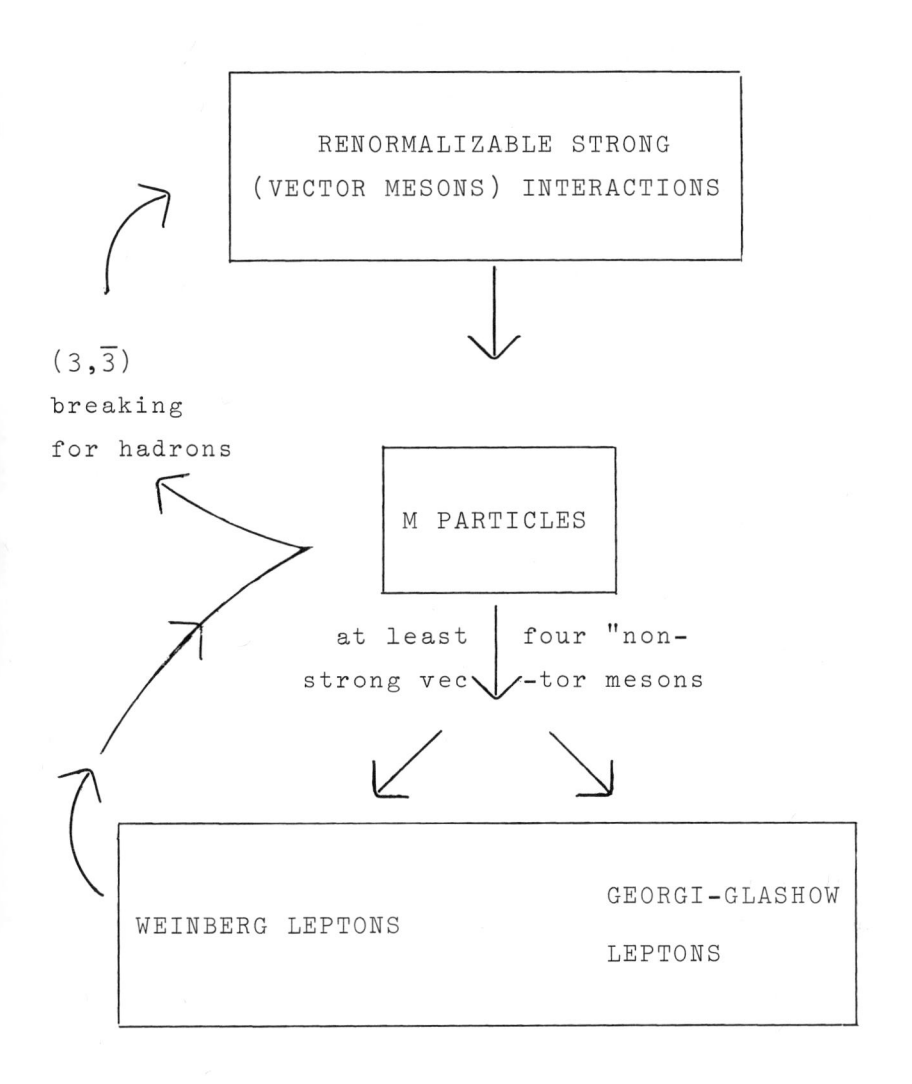

```
                    ┌─────────────────────────────┐
                    │  RENORMALIZABLE STRONG      │
              ↗     │  (VECTOR MESONS) INTERACTIONS│
             /      └─────────────────────────────┘
            /                    │
(3,3̄)      /                     ↓
breaking  /
for hadrons         ┌─────────────────┐
      ↖            │   M PARTICLES    │
        \          └─────────────────┘
         \         at least │ four "non-
          \        strong vec↓ -tor mesons
           ↙                ↘
     ┌──────────────────────────────────┐
     │                    GEORGI-GLASHOW │
     │ WEINBERG LEPTONS                  │
     │                     LEPTONS       │
     └──────────────────────────────────┘
```

(Fig. 1)

FLOW CHART FOR GAUGE THEORY OF ELEMENTARY PARTICLES

leptonic mass patterns. Weinberg's ϕ mixes all four non-strong bosons. In this way, we can suppress neutrino processes in the manner of Georgi and Glashow.

IV. RESUME

We want to summarize our discussion in terms of a flow chart (Fig. 1). The arrows indicate lines of induction in our discussion. At every stage, the new M mesons have played a significant role. They give mass to hadronic vector mesons, couple strong to non-strong interactions, and help generate the physical $3,\bar{3}$ symmetry breaking for the hadrons.

V. RECENT DEVELOPMENTS AND DIRECTIONS

<u>Observability of M's</u>. The masses of the M's are essentially arbitrary in our model. I would like to conjecture, however, that they are in fact quite low mass hadrons. This is indicated in the corresponding dual models. (It would be interesting to put some further high energy (Regge-like) restrictions on the gauge model to see if M masses are predicted.) With I. Bars and K. Bardakci, we have looked at the M-decays. In the first place the nine scalar M's have the quantum numbers of the ordinary σ's, mix with them, and will be hard to observe. The nine pseudoscalars (say P_α) however have abnormal charge conjugation parity. For the singlet, we have $I^G J^P(C) = 0^-0^-(-)$, and for the octet $1^+0^-(-)$. Such states are very unusual and may be hard to observe. In the first place, one notices that these are just the quantum num-

bers of daughters of the B(uddha)(1220) and the h(?) mesons. Thus, I would first hope to find them under the B. P_α has no baryonic decay modes at all (they are 4-quark "exotics" in that language), but can decay into $K\bar{K}\pi$ or five pions. If the P's are under the Buddha they would be <u>very</u> <u>narrow</u> in $K\bar{K}\pi$.

<u>Scaling</u>. With I. Bars and D. Levy, we have investigated the scaling properties in perturbation theory of these models. The currents provide a "new-field-current-identity:" $J^\mu = f(M + k)V^\mu(M + k)$, such that the current attains asymptotic dimension 3. This is reflected in the Feynman graph structure when 2M-intermediate states enter the picture and help provide a nonvanishing W_1 (and a finite νW_2). The Callan-Gross result for "old" algebra of fields is swept aside, but the models we've examined seem no better than the gluon model, in perturbation theory (logs do appear). Still, this makes room for some thought about what electroproduction tells us about constituents: These "constituents" are ρ's and M's (at least in the currents). My own feeling is that a new "phenomenology" can be built around this "new" algebra of fields which will provide an alternative to spin 1/2 partons.

<u>Finite E-M Mass Differences and Parameter Reduction</u>. As we've discussed the SU(3) ⊗ SU(3) model, $\Delta I = 1$ mass differences are not calculable, though they can be fit, using $<\sigma_3>$ as in the tadpole model. There are however, ways of fixing this in the model, but we do not consider them attractive: They involve mathematical, not particularly physical, extra "ingredients", inserted <u>just</u> to get finite e-m mass differences. Since there is nothing more wrong than this, I will list them. (1) Bars and I noticed that if we add <u>another</u> set of three right-right-handed W's coupling only to M's, we prevent any isospin-

breaking insertions, and $<\sigma_3> = 0$ in lowest order. (2) Bars has also noticed the utility of a so-called U group - being another SU(2) which commutes with the basic SU(2) ⊗ U(1) as embedded in U(4) x U(4). Requiring soft (dimension two) breaking of a global U invariance eliminates a number of arbitrary parameters. In fact, we noticed that if Weinberg's φ is a hadron, and parity violation is by hand weak, then the soft U-breaking is enough to fix e-m mass-differences without the extra W's.

REFERENCES

1. K. Bardakci and M. B. Halpern, Phys. Rev. D6, 696 (1972).
2. I. Bars, M. B. Halpern, and M. Yoshimura. Phys. Rev. Letters 29, 969 (1972), and Phys. Rev. (to be published).
3. B. W. Lee and B. de Wit have independently entertained some of these ideas.
4. S. Weinberg, Phys. Rev. D5, 1962 (1972) and 1412 (1972). H. Georgi and S. Glashow, Phys. Rev. Letters 28, 1494 (1972). B. W. Lee, Phys. Rev. D6, 1188 (1972). J. Prentki and B. Zumino (to be published).
5. C. N. Yang and R. L. Mills, Phys. Rev. 96, 191 (1954).
6. S. Weinberg, Phys. Rev. Letters 19, 1264 (1967). A. Salam in Elementary Particle Physics, edited by N. Svartholm (Almquist and Wihsells, Stockholm, 1968), p. 367.
7. A. Neven and J. Scherk (to be published). A. Neven and J. Gervais (to be published).

8. M. Gell-Mann, M. L. Goldberger, F. E. Low, E. Marx and F. Zachariason, Phys. Rev. <u>133</u>, B145 (1964).
9. I. Bars, M. B. Halpern and D. Levy (to be published).
10. P. W. Higgs Phys. Letters <u>12</u>, 132 (1964); Phys. Rev. Letters 13, 508, (1964). T. W. B. Kibble, Phys. Rev. <u>155</u>, 1554 (1967).

CONCERNING THE FORMULATION OF CHARGE INDEPENDENCE OF THE PION NUCLEON INTERACTION*

G. Breit

State University of New York at Buffalo

I. INTRODUCTION

Isotopic spin space (isospace) has been introduced by Heisenberg for stating his hypothesis of charge symmetry of nucleon-nucleon (N-N) interactions. The fruitfulness of the extension to charge independence is apparent through the classic work of Wigner on nuclear structure. The later discovery of the pion (π) combined with Yukawa's famous meson exchange idea made it natural to attribute the approximate charge independence (CI) of N-N interactions to a symmetry of meson-nucleon and in particular pion-nucleon (π-N) interactions in isospace[1] and to regard the physically approximate CI of the N-N system as a consequence of basic interaction symmetries involving meson-nucleon exchanges. One of the main objects of this paper is to

---

*Work supported by the U.S. Atomic Energy Commission (COO-3475-3).

bring out the desirability of a clear and unambiguous formulation of what is meant by CI of the π-N interaction. Since the pion is the lightest meson that can affect the N-N interaction in a major way, a satisfactory formulation of CI for the π-N interaction is also necessary for an understanding of the long-range charge independence (LRCI) of the N-N interaction. Inasmuch as the oldest "unitary symmetries" are the two just mentioned and in view of the importance of unitary symmetries in the classification of elementary particles, it is desirable to formulate clearly what is meant by CI of the π-N interaction. A special aspect of this question has been discussed[2] in connection with speculative estimates of effects of the mass difference $m_{\pi^c} - m_{\pi^o}$ of the charged and neutral pion.

To simplify the immediate discussion, the relatively small dynamic effects of electric charges and magnetic moments on the relative motion of pions and nucleons will be disregarded. But even so the concept of CI of the π-N interaction requires more careful treatment than the usual.

To someone versed in modern "formalisms" it might appear sufficient to use a relativistic charge-independent field-theoretical Lagrangian, to go through the renormalizations, to calculate relevant expectation values and to compare them with experiment. However, renormalization procedures depend on usual power series expansions of the final scattering amplitudes. There are arguments[3] for believing that the perturbation series does not converge. The resultant S amplitude remains unknown therefore. A definite meaning could perhaps be assigned to the series if it were "summable" in some known sense or if some other mathematical meaning could be assigned to it with good reason. This apparently has not been done. Even if a mathematically consistent summation rule should be found it would be necessary to show by comparison with experiment that it gives a good enough representation of fact to make its application as a measurement predictor hopeful. The N-N and π-N interactions problem appears to be much more difficult than that in quantum electrodynamics for which no answers to these cases are available. The π-N interaction constant analogous to the fine structure constant $\alpha \simeq 1/137$ is the $(g^2)_{pv} \approx 0.08$ ("pv" for pseudovector). The expansion parameter is thus roughly 10 times larger. These facts and the associated relatively small hope of finding a convergent modification of the renormalized perturbation series are <u>also generally realized</u>. Furthermore for the N-N interaction the consideration of the π-N interaction is not sufficient because of the nucleon coupling to vector mesons and of the π-π interaction. These in turn affect the π-N interaction through nucleon-antinucleon (N-$\bar{\text{N}}$) pair formation. These matters have been partially discussed previously[4] together with the complication introduced by nucleon structure, the related fact that the neutron mass is greater than that of the proton and the effect of changing the pion-nucleon interaction Ansatz from pseudoscalar (ps) to pseudovector. The problem does not lend itself to subdivision into parts since any one of the interactions is associated with at least one other.

The views expressed above appear to emphasize rigor and logic unduly. One might argue that if all of theoretical physics were pursued without willingness to speculate it would become sterile, that if in the initial development of matrix and wave mechanics one insisted on relativistic invariance of the equations there might not have been available now either the non-relativistic formulation nor such advances in relativistic quantum mechanics as the Dirac electron equation or the Pauli-Weisskopf boson treatment. However, attempts at relativistic generalizations started soon after the matrix-wave equation approach was formulated, Schroedinger, Klein and Gordon having provided the initial efforts and the more decisive steps having followed soon after. Furthermore the Heisenberg-Schroedinger-Born-Dirac nonrelativistic development (QM in brief) replaced the earlier theory of Bohr's Correspondence Principle days and was much superior to it. The empirically arrived at rules which became understandable through the QM such as the replacement of $J^2-L^2-S^2$ by $J(J+1)-L(L+1)-S(S+1)$ were simple, concrete and convincing.

The field-theoretical divergence troubles may disappear as a result of advances in the unification of theories of strong, electromagnetic and weak interactions discussed by Steven Weinberg and others at this conference. It is clear that the subject is still in its formative stages, that the answers to some simple questions are not unique and that perturbation-theory expansions are still there. The advance is too much in the stage of a programme to give an answer to the question of what one can or, as some would say, what one should mean by CI of the π-N interaction.

Most treatments of the π-N interaction used in calculations of the N-N interaction begin with the <u>consideration of the elementary process of pion emission by a nucleon and its inverse.</u> Conservation of momentum obtains in both in usual field theories. For the process and its inverse it reads

$$\vec{p}_N^{\,o} = \vec{p}_{N'} + \vec{p}_\pi \qquad (1)$$

with superscript ^{o} referring to the disintegrating nucleon N which becomes N' on disintegration. The equation still applies after renormalization. There would be no question regarding the evaluation of effects on $\underline{S}$ corresponding to Eq. (1) if it were not for doubts regarding the legitimacy of the evaluation by adding the contributions to the perturbation series of collections of terms $S^{(2)}$, $S^{(4)}$, $S^{(6)}$, where $S^{(n)}$ represents the sum of contributions of $\underline{n}$-pion lines diagrams rather than as

$$S = S'^{(2)} + S'^{(4)} + S'^{(6)} + \ldots, \qquad (2)$$

where

$$S'^{(n)} = \sum_p a^{np} S^{(p)} \qquad (n, p \text{ even}). \qquad (2')$$

Truncation of the series rearranged as in Eq. (2) at $n = 2$ may leave a part of $S^{(2)}$ unaccounted for because it enters $S'^{(4)}$, $S'^{(6)}$, etc. The division of $\underline{S}$ into a part concerned exclusively with OPE and a remainder that is not may thus be impossible to reconcile with singling out of experimental data effects proportional to different powers of the coupling constant $\underline{g}$, some effects of g^2 being present in the omitted parts of $\underline{S}$.

The above caution applies especially to contributions to $\underline{S}$ from low $\underline{L}$ and high $\underline{E}$ from close collisions. Mesons with masses higher than m_π can have an important effect in such cases. Effective potentials obtained by adjustment to phase shifts derived from N-N scattering measurements doubtlessly represent some of the effects of such mesons for a complete field-theoretical treatment of which a generalization of Eq. (2) would be needed. In the absence of such a treatment the evaluation of the pion-nucleon coupling constants $g_{\pi N}$ for different π, i.e. π^c and π^o, and different N, i.e., n and p, cannot be made completely reliably unless the partial waves used for the determination of the $g_{\pi N}$ are free of effects of the heavier mesons, i.e., unless the partial waves employed correspond to high enough L and low enough E. An examination[5] of the contributions to the OPE effect by means of sensitivity curves of that reference or similar considerations can give partial assurance of freedom from complications arising from the a^{np} of Eq. (2') with n ≠ p.

It will be noted that the probable necessity of rearrangement of the perturbation theory series makes employment of the δ-function term (the "contact term") of the OPEP questionable since it belongs to L = 0, for which the direct applicability of the OPEP is not justifiable.

The employment of phenomenological potentials such as the hard-core Yale and Hamada-Johnston or the soft-core Reid potentials in conjunction with the OPE or OPEP remains unjustified and speculative.[6] They are used in the Yale-Buffalo OPE work but only for estimating small corrections such as those caused by nucleon magnetic moments. The effects for relatively high L (but not high enough for inclusion in the OPE group) are of special interest. The objections to the use of static, velocity-independent potentials, while present, are clearly milder in this case than for applications to collisions in general.

The hypothesis that the perturbation-theory series requires at most a rearrangement such as in Eqs. (2) and (2') is also speculative. Its adoption is a plausible though not rigorously carried out justification of representing the series value by its first term in applications to high $\underline{L}$ and low $\underline{E}$ (distant collisions). Analyzing the $S^{(n)}$ and $S'^{(n)}$ in usual spherical harmonics of order $\underline{L}$ normalized to unity and denoting the results by $S_L^{(n)}$, $S_L'^{(n)}$, Eqs. (2) and (2') retain their form. For n = 2 Eq. (2') gives then $S_L^{(2)}$ in terms of $S'_L^{(2\tau)}$, ($\tau \geq 1$). For high $\underline{L}$ and low $\underline{E}$ the $S'^{(2\tau)}$ may be shown to decrease rapidly with $\underline{\tau}$, because the classical mechanics distance of closest approach

$$r_L \approx L/k \ , \ k = Mv/2\hbar \tag{3}$$

$$v = \text{classical mechanics relative velocity} \tag{3'}$$

is large for sufficiently large r_o which condition obtains for large $\underline{L}$ and small $\underline{E}$, the associated value of

$$\exp.(-x_L) \ , \ x_L = m_\pi c r_L/\hbar \tag{4}$$

being small. The influence of the exponential factor of (4) on the a_L^{np} in the $\underline{L}$ component of Eq. (2') is to decrease the importance of the higher $\underline{p}$ for high $\underline{L}$, because for $r < r_L$ the radial function rapidly becomes small on account of the centrigugal barrier. By increasing $\underline{L}$ sufficiently the ratios $S^{(4)}/S^{(2)}$, $S^{(6)}/S^{(2)}$ can be made as small as desired. In this sense $S_L^{(2)}$ can be made a sufficiently good approximation to $S'_L^{(2)}$.

The mathematical proofs of the statements in this paragraph have not been completed but there is little doubt of their validity.

A qualitative test of validity of the conclusions can be made by the examination of the self-consistency of consequences of the assumption about the sufficiency of the rearrangement such as in Eqs. (2) and (2'). For example, E_{max}, the maximum energy of experimental data used in the determination of $g_{\pi N}$ can be varied and the stability of $g_{\pi N}$ examined. The changes produced are in the distance of closest approach between the nucleons admitted in the analysis. Various consistency tests connected with this point have been made by the Yale-Buffalo group. No definite contradiction has been found so far.

An important need for modification could and probably has escaped detection, however. This could be caused by the interaction of a pion emitted by nucleon N with the cloud of other pions (the π-π interaction) and other mesons reaching a similar cloud around N' and finally N' itself. It might appear at first sight that such an error could be ruled out making use of the probable permanence of the positions and constitutions of these clouds with respect to N and N'. Were the π emitted by N prevented from reaching N' by absorption within the cloud belonging to N, the constitution of that cloud would change contrary to the temporary assumption made. But this assumption is reasonable only if the collision is adiabatic. In the classical-mechanics picture, however, there is no binding reason for supposing that the pion-meson cloud of N at one point of the orbit in the relative-coordinate space of N and N' is the same as at another point or that it is oriented in the same way with respect to the line joining $\underline{N}$ to N'. Even though adiabaticity of the collision is probable for very high $\underline{L}$ the validity of particle conservation is not at variance with the possibility that the pion is scattered elastically by parts (or the whole) of the pion clouds surrounding $\underline{N}$ and N' before the pion is absorbed by N'. The possibility exists therefore that somewhat different meanings of the usual $g_{\pi N}^2$ apply to different determinations, and that such agreement as there is between the Samaranayake and Woolcock[7] ($g_o^2 = 15.9 \pm 0.3$ for p-p) value of g^2 which makes no use of N-N scattering data and values from the latter is partly fortuitous.

Another view of the meaning of g^2 as obtained from distant collisions in N-N scattering is obtained from the OPEP picture. Neglecting relativistic corrections which can be worked out[5] by an additional calculation, the usual OPEP can be obtained[8,9] from the energy change caused by bringing $\underline{N}$ and N' together, in static approximation, and subtracting the self-energy parts of the pion field. The cross-product term responsible for the OPEP is affected by the scattering of the pion by the cloud as in the consideration of the preceding paragraph. The OPEP picture thus supports the conclusion that there are in principle two coupling constants. One of them (the first) is that applying to the elementary process to which the usual Eq. (1) is supposed to be applied after the usual renormalization but neglecting the clouds. The other describes the action of the whole completely dressed nucleon including the effects of the pions, the ω, ρ and other mesons. The first kind of pion exchange appears difficult to isolate for measurement.

Were one to attempt the formulation of a definition of $g_{\pi N}$ purely phenomenologically say in terms of measuring the OPEP one would be led to the second of the two kinds of $g_{\pi N}$, i.e., the definition applying to N-N scattering but not that needed for developing a field theory of the interaction in the usual order.

The existence of a difference between the two kinds of OPE is detectable in principle. For example, if the angular distribution of the scattering amplitude caused by OPE could be tested employing N-N scattering experiments the deviations from expectation from the well-known theory for OPE #1 would support the view that $(OPE)_1 - (OPE)_2 \neq 0$. The subscripts 1 and 2 indicate here whether the first or the second kind of OPE is meant. Since the determination of the whole angular distribution is impossible with available accuracy of scattering data a partial but more practical answer to attempt might be obtainable by inserting form-factor like modifications of $(OPE)_1$ amplitudes calculated from plausible models of the meson clouds and to determine "best" values (in a statistical sense) of one or two adjustable model parameters by comparison with relevant available N-N scattering data. Although tests of the mathematical form of the OPEP have been made[10] the accuracy of the data collection used would have been insufficient for the present purpose, nor could a definite promise of success be made with the more complete data and improved methods of calculation now available.

II. ROLE OF PION MASS DIFFERENCE

The long-range N-N interactions (LRCI) caused by

π^o and $\pi^c = \pi^{\pm}$ exchanges vary differently with the internucleon separation distance $\underline{R}$ for p-p and n-p interactions. If the proportionality constant in the OPE formula is adjusted to give equality of the interactions at one $\underline{R}$ it will not give it at another.[11] Similarly if the pion field is adjusted by a change of the coupling constants to be the same for π^o and π^c at a distance $\underline{r}_{\pi N}$ between pion and nucleon it will not be the same at another $\underline{r}_{\pi N}$. These differences depend on whether the pseudoscalar (ps) or pseudovector (pv) pion-nucleon interaction theory is used on account of the extra m_π^2 in the OPE and the OPEP.

The dominant factor for the dependence of OPE phase shifts on $\underline{R}$ is

$$e^{-x}, \quad x = (m_\pi c/\hbar)R \quad . \tag{5}$$

In some of the determinations of the pion-nucleon coupling constant g^2 from N-N scattering the collisions having an important effect for orbital angular momentum quantum number $L = 5$ (i.e. for $\underline{H}$ waves) correspond to incident nucleon energy 180 MeV, which on the classical turning point criterion, gives for the smallest $\underline{R} = 3.72$ F for which $x = 2.58$. According to Eq. (1) a change of m_π by 3.4% changes the smallest important $\underline{x}$ by 0.094 which affects e^{-x} by 9.4%. At values of $\underline{R}$ greater than the minimum important one and at energies > 180 MeV, the change in e^{-x} is even greater. On the other hand, the values of g^2, g^2_{p-p} and $(g^2)_{n-p}$ from p-p and n-p data, after various corrections, appear to agree with each other appreciably better than that. In the determination of $(g^2)_{n-p}$ the OPE phase shifts δ considered as functions of m_π enter in the combinations

$$2\delta(m_{\pi^c}) - \delta(m_{\pi^o}) \quad (I = 1)$$

$$\frac{1}{3}[2\delta(m_{\pi^c}) + \delta(m_{\pi^o})] \quad (I = 0)$$

(6)

for the two possible values of the isospin I. For I = 1 the presence of a difference in absolute values increases the sensitivity to the effect of $m_{\pi^c} - m_{\pi^o}$. It is thus necessary to have the theory of the OPE interaction good enough to provide the effect of $m_{\pi^c} - m_{\pi^o}$ on the δ.

The question of formulating the π-N interaction <u>in</u> a <u>form</u> identical to that of the ps or pv theories will be considered next with the limited objective of securing the correct dependence of the pion field produced by a nucleon asymptotically at large $r_{\pi N}$. Effects of nucleon structure, nucleon and pion sizes depending on $r_{\pi N}$ are thus largely eliminated. This question is inseparable from that of the origin of the mass difference $m_{\pi^c} - m_{\pi^o}$, for clearly if this difference is the result of a difference in the internal constitutions of π^c and π^o then it is unreasonable to use the same interaction Hamiltonian for both of them. A qualitative indication regarding the answer is obtainable from the Coulomb self-energy of m_{π^c}. Omitting largely irrelevant, model-dependent numerical factors, and taking for $m_{\pi^c} - m_{\pi^o} = 4.57$ MeV as e^2/r, gives

$$r = 0.32 \text{ F}.$$

Here <u>r</u> is a mean distance between elements of charge of the actual charge distribution in its assembly from an infinite distance which for a surface charge distribution over a sphere of radius <u>a</u> is 2<u>a</u> but for a uniform volume distribution inside a sphere of the same radius is appreciably smaller leading to a larger Coulomb energy. The value 0.32 F may be compared with

$$\hbar/M_N c = 0.21 \text{ F} \qquad (M_N = \text{nucleon mass})$$

which is a measure of the distance through which a nucleon wave packet can be concentrated employing free plane waves of positive energy only at smaller distances N-$\bar{N}$ pairs interfering with pion localization. Since the two characteristic lengths are comparable, the pion model may be considered as having a bearing on reality. A theory in which the electrostatic energy of the pion charge distribution is simply added to the energy of the field is thus not out of the question.

An objection to this view is that the main features of pion-field theory would remain somewhat artificial. In it a theory of charged fields is first developed. The two charged pions appear on an equal footing and the equality of their masses follows from one being the antiparticle of the other, a connection rooted in the fact that repeating complex conjugation gives the original charge back again. To the field, the properties of which depend on such a fundamental relationship, there is then added the field of the neutral pion. The π^o and π^c masses are postulated to be equal and the Hamiltonian is postulated to be invariant under isospin-space orthogonal transformations. The latter step involves identifications of coupling constants of the two pion-field ingredients as follows

$$g_p = -g_n = g_c/\sqrt{2} = g \quad . \tag{7}$$

Here $\underline{p}$, $\underline{n}$, and $\underline{c}$ stand for proton, neutron and charged respectively; g_p and g_n are the values of the pseudoscalar coupling constant for $\underline{p}$ and $\underline{n}$ when coupled to π^o and g_c is the value for the charged field. The minus sign insures the coupling term to π^o having a different sign when a proton is replaced by a neutron. These requirements arise from the desire to secure agreement with Kemmer's form[1] of the π-N interaction, which gives results invariant to rotations in isospin space and consequently secures the same for N-N interactions. Eq. (7) is no more than a statement of partially established relationships suggested by experimental evidence concerning N-N interactions. Nevertheless one tends to prefer a theory securing fundamental symmetries <u>in virtue of its primary structure</u> rather than through additional conditions. The inequality of m_{π_c} and m_{π_o} detracts from faith in the fundamental character of Eq. (7). It is hard to believe, therefore, that the approximate estimates discussed earlier <u>cover</u> the situation even though it is likely that they have much to do with it and that without allowance for the electrostatic effects on the effective pion mass any consideration of charge independence (CI) is incomplete.

The Fermi-Yang (FYM) model of the pion received little attention but it is probably a <u>qualitative</u> and essentially correct picture. It has since become clear that some interactions between nucleons and vector mesons are large and that one is back at the FYM in a modernized form. Whether the "undressed" pion of field theory is believed in or not, the "dressed" pion probably contains an appreciable component of the N-$\bar{\text{N}}$ type. The situation is similar to e^+-e^- pair production around an electric charge giving vacuum polarization. The

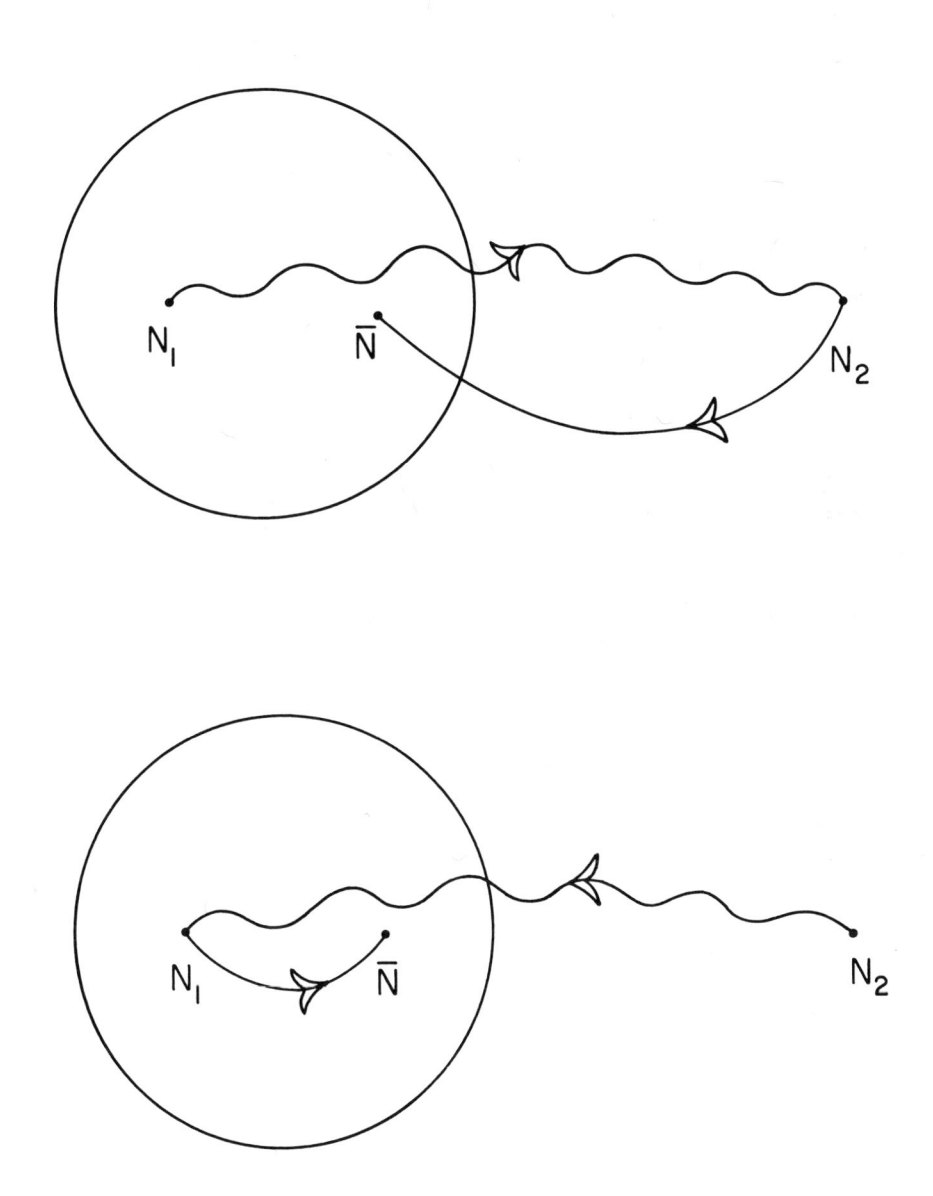

Fig. 1. Wavy lines for propagation of influence of vector-meson field, full lines for jump from E > 0 state to hole among Dirac sea of occupied E < 0 states.

importance of the elementary π-N interaction is thus subordinated by the appreciable density of N-$\bar{\text{N}}$ pairs whenever the pion is localized in a volume of a radius comparable with $\hbar/M_N c$. The FYM thus describes an appreciable component of the pion wave function whether the bare pion is replaced by its FYM picture or not.

Considering for the sake of concreteness the neutral vector meson, the π-N absorption may be pictured[12] in two alternative ways as in the two parts of Fig. 1.

In that figure wavy lines represent the influence of the vector meson field, the arrow pointing from the wave source to the nucleon acted on. Full smooth lines represent the transition from a positive-energy level to a hole in the Dirac Sea of negative energy states normally occupied by nucleons. The presence of the hole is indicated by $\bar{N}$. The upper part of the figure pictures a vector meson field component emanating from a positive energy nucleon N_1 inside the pion, inducing a transition from a positive energy nucleon N_2 into the hole associated with $\bar{N}$ and thus causing recombination of nucleon N_2 with π. The lower part of the figure shows a vector meson field emanating from nucleon N_2 reaching positive-energy nucleon N_1, inducing a transition from a state of N_1 with $E > 0$ to a vacant negative-energy state interpreted as antinucleon $\bar{N}$. The upper part represents the recombination of π with nucleon outside of it; the lower, the disappearance of a pion because it is exposed to a nucleon that remains outside of it. Heisenberg's uncertainty principle interferes of course with a literal acceptance of space locations of participating particles.

On the FYM the two charged pions are represented

similarly, <u>viz.</u>, π^+ as $p+\bar{n}$, π^- as $\bar{p}+n$, but for π^0 there are available both $p+\bar{p}$ and $n+\bar{n}$. The masses of π^+ and π^- should therefore be equal but the mass of π^0 may be different because of coupling of $p+\bar{p}$ to $n+\bar{n}$. Even disregarding the electrostatic attraction between p and $\bar{p}$, the ground state of the level formed has to be lower than those of the constituents in agreement with m_{π^0} and m_{π^c}. That the difference $m_{\pi^c} - m_{\pi^0}$ is relatively small indicates that the lowering of the energy caused by degeneracy is small and agrees with $\underline{g}$ being nearly the same for π^0+N and π^c+N. There are questions to clear up regarding this model but it is doubtlessly promising.

Starting with assigned coupling constants for ω-N the value of $g_{\pi-N}$ can be estimated for an assigned radius of the pion model. The latter however implies a value of the N-$\bar{N}$ interaction which in turn implies a value of the N-N, and therefore of the π-N interaction and hence gives a condition on $g_{\pi-N}$. Self-consistency of the whole model thus results in an approximate value of the π-N interaction. Crude estimates indicate approximate agreement with observation.[12]

The immediately preceding discussion so far has been concerned exclusively with the possibility that the pion wave function is in some sense a linear combination of N-$\bar{N}$ states. If isospin is strictly conserved as a consequence of charge independence assumed also for all interactions involving vector mesons used in the model, the triplet of states with $I_\zeta = 1, 0, -1$ remains isolated from states with values of the total isospin I other than $I = 1$. A complete isolation is prevented by electromagnetic interactions and probably by other effects. Perturbations by $I = 0$ cannot affect the

$I_\zeta = \pm 1$ states, but non-negligible effects on π^o ($I_\zeta = 0$) may have to be considered. The η meson (m = 549 MeV) should thus be included in the consideration. It is believed to have a small but nonvanishing η-N coupling constant g_N and enters the SU_3 classification. The isospin impurity of the physical π^o state resulting from this coupling may be expected to lead to π^o-N coupling being shared with η^o-N coupling. In a strict sense therefore the problem requires dealing with more extended spaces than that of the isospin.

It will be recalled that Fermi and Yang (employing a somewhat unacceptable interaction potential) arrived at a linear combination of ps and pv couplings for the π-N interaction. Although an important cornerstone of their deduction is shaky, their general <u>conclusion</u> appears <u>sound</u> and <u>probable</u> as seen from Fig. 1. It shows that the geometrical size of the pion matters for the π-N interaction mechanism. If it is assumed that the interaction is ps then it appears likely that a pv part of the interaction exists also, a Taylor expansion in powers of the pion radius being needed for describing the processes pictured in the figure. The physical pion is not the point particle to which the elegant enumerations of invariants and renormalizability arguments may be applied.

There are thus at least three possible causes, each of which may lead to different mass values of π^c and π^o as well as of $g^2_{\pi^o N}$ and $g^2_{\pi^c N}$: (a) the ordinary Coulomb interaction; (b) near degeneracy of $1^{\underline{st}}$ - order representation of π^o in terms of FYM wave functions leading to large sensitivity to perturbations; (c) violation of conservation of <u>I</u> caused by electromagnetic effects and by the complete classification of states in terms of quantum numbers requiring more than the assumption of strong-interaction invariance under orthogonal transformations of the isospin space (Heisenberg, Wigner).

The treatment of electromagnetic effects as superposable on the strong interactions effects by additions to the Lagrangian may prove to be no more than a convenient but only approximately valid working hypothesis. It is doubtful, however, that the main phenomenological foundations of this paper need <u>as yet</u> the consideration of such revisions in fundamental physical approaches. Attention should be called <u>once more</u> to the artificiality of achieving complete charge independence of the pseudoscalar meson field with nucleons being removed above on the assumption that for the vector meson fields a similar charge independence is assumed to hold well enough, a hypothesis convenient for N-N interaction theory.

III. RECENT EVIDENCE FROM DATA ANALYSIS

Since the move of the Yale N-N scattering-data analysis group to SUNY at Buffalo a number of improvements in the comparison of g^2 values for $(\pi^o-p)_{pp}$, $(\pi^c-n)_{np}$, $(\pi-N)_{np}$ have been made[13] regarding enforcement of continuity at the transition energies, inclusion of corrections for two-pion exchange, nucleon magnetic moment corrections, separation of π^c from π^o effects in the case of n-p scattering, and changes in upper limit of incident nucleon energy of experimental data used, so as to eliminate as much as possible effects of inelasticity of N-N collisions. No statistically significant differences between $(g^2)_{p-p}$ and

$(g^2)_{n-p}$ have been found so far.

The general method used in the Yale-Buffalo work consists in treating phase shifts of N-N partial waves of high orbital angular momentum $\underline{L}$ at the lower energies as caused by one-pion exchange and in adjusting g^2 to give the best fit to the data.[14]

In obtaining the results quoted below, pure ps coupling was used somewhat arbitrarily even though the arguments at the end of the preceding section speak against this being the case. The experimentally known pion masses were used literally in all of these calculations. In both instances this was largely a concession to need for budgetary economy and for simplicity of work records.

Although the employment of potentials used for some of the values in Table I has its illogical sides, it has the following positive features: (a) it is expected that the magnetic moment corrections (mmc) matter for g^2 comparisons mainly at the larger $\underline{R}$ where the role of potentials is mainly that of reproducing phase shifts on the verge of belonging in the OPE group; (b) the older method, while sound in principle, involves a purely empirical adjustment of correction functions which cannot be carried out with certainty without introducing too many adjustable parameters into the main search procedure to be practical.

The newer results of special interest for the main topics of this paper are summarized in Tables I and II below.

Table I shows the last two stages of improvement employing the data collection Y-IV, (1962 deadline). The TPE corrections used the published work of S.N. Gupta, R.D. Haracz, and collaborators. The old MAG (abbreviation for magnetic moment corrections) did not provide suitable functions for adjustment to data without losing accuracy. The new MAG makes use of calculations with plausible potentials. It also is not invulnerable to criticism but is believed to give an appreciable improvement. The very close agreement of p-p with n-p values is probably fortuitous.

TABLE I

Effects of successive inclusion of TPE* and of change from old to new MAG** procedure on $g_o^2 = g^2/\hbar c$ from $(Y-IV)_{p-p}$ and $(Y-IV)_{n-p}$*** M data.

TPE	MAG	$(g_o^2)_{p-p}$	$(g_o^2)_{n-p}$	$(g_o^2)_{n-p} - (g_o^2)_{p-p}$
No	Old	13.74±0.53	14.70±0.52	0.96±0.74
Yes	Old	15.11±0.62	14.67±0.51	-0.44±0.80
Yes	New	14.59±0.62	14.69±0.51	0.10±0.80

In Table II the effects of extending the data deck by modernizing it and of changing the choice of "correction functions" and especially the procedure in applying them is illustrated together with the effect of decreasing the maximum bombarding energy used. The

---

*Two-pion exchange.
**Nucleon-magnetic-moment corrections.
***The later Yale treatment of p+d data.

latter step is intended to minimize the effect of producing pions in <u>open</u> reaction channels which is difficult to take into effect reliably. The present indications are that the previously arrived at results are essentially correct.

TABLE II

Effects of data collection, maximum data energy and correction functions on the "difference" $[(g_o^2)_{n-p} - (g_o^2)_{p-p}] \pm [(\Delta g_o)^2_{n-p} + (\Delta g_o)^2_{p-p}]^{1/2}$.
No two-pion exchange corrections, no improvements in Yale magnetic moment corrections.

Type of Search		Data Collection	
		Y-IV	Extended
(Y-IV) pattern of correction functions. $(E_{max})_{n-p} = 350$ MeV, $(E_{max})_{p-p} = 345$ MeV	$(g_o^2)_{n-p}$	14.23±0.28	14.91±0.35
	$(g_o^2)_{p-p}$	13.01±0.42	14.20±0.28
	difference	1.22±0.51	0.71±0.45
f_J^L correction functions $(E_{max})_{n-p} = 350$ MeV, $(E_{max})_{p-p} = 345$ MeV	$(g_o^2)_{n-p}$	14.70±0.52	14.71±0.48
	$(g_o^2)_{p-p}$	13.74±0.53	14.42±0.42
	difference	0.96±0.74	0.29±0.64
f_J^L correction functions $(E_{max})_{n-p} = 310$ MeV, $(E_{max})_{p-p} = 316$ MeV	$(g_o^2)_{n-p}$	14.56±0.63	15.05±0.57
	$(g_o^2)_{p-p}$	13.78±0.51	14.45±0.42
	difference	0.78±0.81	0.60±0.71
f_J^L correction functions $(E_{max})_{n-p} = 272$ MeV, $(E_{max})_{p-p} = 260$ MeV	$(g_o^2)_{n-p}$	14.30±0.69	14.45±0.64
	$(g_o^2)_{p-p}$	13.92±0.52	14.68±0.43
	difference	0.38±0.86	-0.23±0.77

Tables I and II together support the assumptions made regarding the sense in which charge independence of the OPE interaction holds. There are, however, several sources of error which have not been examined sufficiently. A final conclusion must be postponed therefore.

It is desired to acknowledge the helpful collaboration of Drs. Tischler and Mukherjee on material for Table I, to Drs. Tischler and Pappas on that for Table II, to the staff of Computer Services of SUNY/B for their helpfulness, and to Mrs. Doris Barker for painstaking manuscript preparation.

Thanks are due to Professor Danielli, Director of the Center of Theoretical Biology, State University of New York at Buffalo, for his hospitality in providing working space.

REFERENCES

1. N. Kemmer, Proc. Cambr. Phil. Soc. <u>34</u>, 354 (1938).
2. G. Breit, Proc. Nat. Acad. Sci. US <u>63</u>, 223 (1969); Nuclear Phys. <u>B 14</u>, 507 (1969).
3. S.S. Schweber, H.A. Bethe, and F. de Hoffman, <u>Mesons and Fields</u> (Row, Peterson and Company, Evanston, Illinois, 1956), Vol. I., and references therein. F. Mandl, <u>Introduction to Quantum Field Theory</u> (Interscience, New York, 1959), pp. 160-161 with their references especially.

    Although some of the more recent work suggests that the possibility of convergence of the perturbation-theory series is not excluded, it appears unlikely because there is experimental evidence of nonvanishing pion-pion interaction and demonstrated lack of convergence in the boson-boson case.

    In view of serious doubt regarding the convergence, the text uses the safe approach of not assuming that convergence of the series is assured.
4. Second paper quoted in Ref. 2
5. G. Breit and M.H. Hull, Jr., Nuclear Phys. <u>15</u>, 216 (1960), Figs. 1 and 2 especially.
6. G. Breit, <u>Proceedings International Conference on Properties of Nuclear States</u>, Montreal, Canada, August, 1969 (Les Presses de L'Université de Montréal), p. 293.
7. V.K. Samaranayake and W.S. Woolcock, Phys. Rev. Letters <u>15</u>, 936 (1955).
8. W. Pauli, <u>Meson Theory of Nuclear Forces</u> (Interscience, New York, 1946).
9. G. Breit, Phys. Rev. <u>111</u>, 652 (1958), Sect. II especially.

Neither this paper nor Pauli's book contain a proof of the equivalence of the unquantized field derivation of the OPEP to that using quantized waves. The equivalence of the two can be readily demonstrated by comparing equations on the amplitudes in the two cases.

10. G. Breit, M.H. Hull, Jr., K. Lassila and H.M. Ruppel, Phys. Rev. Letters 5, 274 (1960); Proc. Nat. Acad. Sci. 46, 1649 (1960).

11. Since from now on confusion might arise between the internucleon separation distance and the distance of a pion from a nucleon the symbol R will be used for the former.

12. G. Breit, Proceedings International Conference on Nucleon Structure (Stanford Univ. Press, Stanford, California, 1964). In the relevant figure of that reference the arrows on the vector meson field lines are reversed.

13. G. Breit, M. Tischler, S. Mukherjee, and J. Lucas, Proc. Nat. Acad. Sci. US 68, 897 (1971); G. Breit, Proceedings Ninth Coral Gables Conference on Fundamental Interactions, January, 1972, Coral Gables, Florida; G. Breit, M. Tischler, and S.N. Mukherjee, On Magnetic Moment Corrections in N-N Scattering (to be published).

14. G. Breit and R.D. Haracz, "Nucleon-Nucleon Scattering" in High Energy Physics (Academic Press Inc., New York, 1967), Vol. I and references therein.

THEORY OF A ONE-DIMENSIONAL RELATIVISTIC ELASTIC
CONTINUUM AND HADRONIC WAVE EQUATION*

Takehiko Takabayasi

Department of Physics

Nagoya University, Nagoya, Japan

After Nambu and others derived the dual amplitude based on a string-like model,[1] we derived the <u>relativistic quantum mechanics</u> of a finite one-dimensional elastic continuum ("string"), taken as a model of hadrons, by a "detailed wave equation".[2] We want here to reconstruct the theory by first giving the classical theory of a relativistic string and then quantizing it,[3] in such a way that the method clarifies some general points which will be applicable to any more general relativistic extended model.[4]

We start from the two parameter representation

$$x_\mu = x_\mu(\sigma, \tau) \qquad (1)$$

for the "world strip" swept by the string's motion, where σ is the label for each element of the string, and

---

*A detailed account of this work is given in a lengthy preprint with the same title (January 12, 1973, Department of Physics, Nagoya University).

$\sigma_1 \leqq \sigma \leqq \sigma_2$, while τ is an "instant parameter". Purely kinematically, however, they are largely arbitrary. By a reparametrization

$$\sigma \to \sigma'(\sigma,\tau), \quad \tau \to \tau'(\sigma,\tau), \tag{2}$$

the same world strip is represented as $x_\mu = x_\mu'(\sigma',\tau')$. Two simple quantities invariant under (2) are the line element lying on the strip $(dx_\mu)^2 = G_{11} d\sigma^2 + G_{00} d\tau^2 + 2 G_{01} d\sigma d\tau$ and the surface element $(G_{01}^2 - G_{00} G_{11})^{1/2} d\sigma d\tau$, where $G_{00} = (\frac{\partial x_\mu}{\partial \tau})^2$, $G_{11} = (\frac{\partial x_\mu}{\partial \sigma})^2$, $G_{01} = \frac{\partial x_\mu}{\partial \sigma} \frac{\partial x^\mu}{\partial \tau}$ represent the metric tensor on the strip, and $G_{01}^2 - G_{00} G_{11} \geqq 0$. Now (1) is a (super) many-time representation because it means

$$x_i = x_i(\sigma,\tau), \quad x_o = x_o(\sigma,\tau). \tag{3a,b}$$

We must be able at any stage of the theory to put all individual times equal to a common time t such that $x_o(\sigma,\tau) = t$, and by solving this and inserting the result into (3a) we get the ordinary (noncovariant) representation $x_i = x_i(\sigma, \tau(\sigma,t)) = x_i(\sigma,t)$ for motion. Thus the four-velocity of an element of the string is $U_\mu = (\frac{1}{\sqrt{-G_{00}}}) \cdot \frac{\partial x_\mu}{\partial \tau}$. The parameter τ is employed for the purpose of covariant representation and is also convenient for exhibiting the conformal symmetry of theory, but our main concern should be the <u>uniqueness of physical interpretation</u> on one hand and the necessity of assuring <u>causality</u> on the other, in such a description which relies on a redundant parameter. In classical theory this necessarily requires that the complete elimination of τ should be possible in leading to a noncovariant but unique description, whereas in quantum theory it requires that τ should completely disappear in the Schrödinger picture.

Now causality demands that $G_{00} \leq 0$ and $G_{11} \geq 0$ everywhere. They are consistent with the condition of choosing (σ,τ) as orthogonal curvilinear coordinates:

$$G_{01} = 0. \qquad (4)$$

The mechanical law is given by the equation of motion and the boundary condition

$$\frac{\partial^2 x_\mu(\sigma,\tau)}{\partial \tau^2} = \frac{\partial^2 x_\mu(\sigma,\tau)}{\partial \sigma^2}, \quad \frac{\partial x_\mu(\sigma,\tau)}{\partial \sigma}\bigg|_{\sigma_1,\sigma_2} = 0, \qquad (5a,b)$$

both of which follow from the variational principle with the Lagrangian density $L = \frac{K}{2}(G_{00} - G_{11})$, and by the "normalization condition"

$$\Lambda^\circ = \frac{K}{2} \int_{\sigma_1}^{\sigma_2} (G_{00} + G_{11}) d\sigma = -\omega_\circ, \quad \omega_\circ \geq 0. \qquad (6)$$

We have then $G_{00} + G_{11} = \text{const} = -\frac{2\omega_\circ}{K(\sigma_2 - \sigma_1)}$. The fundamental equations (4), (5) and (6) are invariant under internal scale transformation, where the scale-invariant constant is $\Omega = 2(\sigma_2 - \sigma_1)\omega_\circ/K$. Furthermore (5a,b) are invariant under conformal transformations but (6) destroys this invariance unless $\Omega = 0$. To write down the general solution for the fundamental equations we note that the whole motion is determined once we are given the <u>motion of the end</u>, $x_\mu(0,\tau) = u_\mu(\tau)$, which must have the property that $\frac{du_\mu(\tau)}{d\tau} = v_\mu(\tau) = \frac{\partial x_\mu(0,\tau)}{\partial \tau}$ satisfies[5]

$$v_\mu(\tau) = v_\mu(\tau + 2\pi), \quad (v_\mu(\tau))^2 = -\Omega/\pi^2. \qquad (7a,b)$$

The solution is simply $x_\mu(\sigma,\tau) = \frac{1}{2}(u_\mu(\tau-\sigma) + u_\mu(\tau+\sigma))$. Also we verify that $(\sqrt{\Omega}/\pi)\tau$ is the <u>proper time of the end</u>. Via $v_\mu(\tau) = \sqrt{2}\sum_{r=-\infty}^{\infty} c_\mu^r e^{-ir\tau}$, $v_\mu(\tau)$ can be replaced by an infinite number of constant vectors c_μ^r, among which $\sqrt{2}\pi\kappa c_\mu^o = P_\mu$ is the four-momentum of the system. Also (7b) is re-expressed as

$$\Lambda^r = -\omega_o \delta_{ro}, \quad (r=0, \pm 1, \pm 2, \ldots) \qquad (8)$$

where the $\Lambda^r = \pi\kappa \sum_n c^n c^{r-n}$ are generators of conformal transformation. Solutions satisfy $m^2 = -(P_\mu)^2 \geq \kappa^2\Omega$ so that no tachyons occur, and the leading trajectory is given by $S = \frac{m^2}{2\pi\kappa} - \frac{\kappa\Omega}{2\pi}$, where S is the length of spin: $S = [-(\frac{1}{2}\varepsilon_{\mu\nu\kappa\lambda}M^{\nu\kappa}P^\lambda)^2/P^2]^{1/2}$. Our set of fundamental equations is shown to be indeed equivalent to non-covariant equations for $x_i(\sigma,t)$ without constraint. In the special case $\Omega=0$ the whole theory is invariant under conformal group, so that the definition of (σ,τ) leaves this arbitrariness, and we can utilize this to impose further constraint allowing transverse oscillations alone.

Now we quantize the theory, in the Heisenberg picture regarding τ, by establishing equal τ commutation relations between $x_\mu(\sigma,\tau)$ and $p_\mu(\sigma,\tau) = \kappa \partial x_\mu(\sigma,\tau)/\partial\tau$. Then for unequal τ we verify that $[x_\mu(\sigma,\tau), x_\nu(\sigma',\tau')] = 0$, for $|\tau-\tau'| < |\sigma-\sigma'|$, which expresses causality on the basis of $G_{11} \geq 0$. Corresponding to the classical constraint (8), we should impose the subsidiary conditions

$$(\Lambda^r + \omega_o \delta_{ro})\Psi_H = 0, \quad (r=0, 1, 2, \ldots) \qquad (9)$$

since only $\{\Lambda^o + \omega_o, \Lambda^r; r > 0\}$ form a closed algebra.

Next we go over to the Schrödinger picture via
$\Psi_S(\tau)=e^{-i\Lambda^o\tau/\hbar} \cdot \Psi_H$. Then $\Psi_S(\tau)$ must be a stationary solution with the fixed frequency $-\omega_o$ in τ, and $\Psi_S(0)\equiv\Psi$ itself must satisfy $(\Lambda^r+\omega_o\delta_{ro})\Psi=0$, $r=0, 1, 2,\ldots$, which is re-expressed precisely as our original detailed wave equation $H(\sigma)\Psi[x(\sigma)]=0$ with $H(\sigma)=\frac{\kappa}{2}(G_{00}+G_{11})+\frac{i\kappa}{\pi}\int_o^\pi G_{01}(\sigma')\frac{\sin\sigma'}{\cos\sigma-\cos\sigma'}d\sigma'+\frac{\omega_o}{\pi}$, and τ disappears completely. The absence of tachionic states in classical theory corresponds to the absence of negative norm states in quantum mechanics when one understands indefinite-metric Hilbert space.

To consider the case with interaction we introduce the coupling with an external scalar field $\phi(x)$ at the end, represented by the interaction Lagrangian density $L'=-(2g/\pi)\delta(\sigma)\phi(x(0,\tau))$. In classical theory the constraint now generally involves quantities at unequal τ, whence τ cannot be eliminated, except in the case where the external force is Minkowskian: $U_\mu(0,\tau)[\partial\phi(x(0,\tau))/\partial x_\mu(0,\tau)]=0$. On the other hand in quantum mechanics the elimination of τ in the Schrödinger picture requires that the external particle mass should be "quantized" to the unphysical Virasoro value $m_{ex}^2=-2\pi\hbar\kappa$, and in spite of the original conformal non-invariance of the external interaction, the conformal symmetry is partially restored as gauge symmetry. Again the fundamental equation is represented by a detailed wave equation, which is obtained by adding $-L'$ to $H(\sigma)$.

REFERENCES

1. Y. Nambu, Proceedings of International Conference on Symmetries and Quark Models (1969), p. 269. L. Susskind, Nuovo Cim. <u>69A</u> (1970), 457. T. Takabayasi, Prog. Theor. Phys. <u>43</u> (1970), 1117.
2. T. Takabayasi, Prog. Theor. Phys. <u>44</u> (1970), 1429; <u>46</u> (1971), 1528 and 1924.
3. Such reconstruction has been investigated by several authors: O. Hara, Prog. Theor. Phys. <u>46</u> (1971), 1549. T. Goto, Prog. Theor. Phys. <u>46</u> (1971), 1560. F. Mansouri and Y. Nambu, Phys. Letters <u>39B</u> (1972), 375. G. Konisi, Prog. Theor. Phys. <u>48</u> (1972), 2008.
4. Note that our simple string model can be generalized by endowing each elementary constituent with internal "Fermi oscillator" behavior (i.e. pairs of Dirac matrices) or parafermi behavior. See T. Takabayasi, Prog. Theor. Phys. <u>47</u> (1972), 1026; <u>48</u> (1972) 1718.
5. Here we have fixed $[\sigma_1, \sigma_2]$ as $[0, \pi]$. By doing this the manifest scale invariance becomes invisible, but we fix this below.

AN APPROACH TO HADRON PHYSICS THROUGH ASYMPTOTIC SU(3)
AND THE CHIRAL SU(3) x SU(3) ALGEBRA[*]

S. Oneda[**]

Center for Theoretical Physics

University of Maryland, College Park, Md. 20742

and

Seisaku Matsuda

Department of Physics

Polytechnic Institute of Brooklyn

Brooklyn, New York 11201

We study in a systematic way, the inter-SU(3)-multiplet regularities and clues to the possible level scheme of hadrons.

We cope with <u>broken</u> SU(3) by using the hypothesis[1] of asymptotic SU(3) which can be made in the presence of G-M-O mass splittings with mixing.

With our asymptotic SU(3), we extract information solely from the chiral SU(3) x SU(3) charge algebras[2] and also from the <u>exotic</u> C.R.'s, such as $[\dot{V}_{K^0}, V_{K^0}] = [\dot{V}_{K^0}, A_{\pi^-}] = [\dot{V}_{K^0}, A_{K^0}] = 0$ $(\dot{V}_{K^0} = \frac{d}{dt} V_{K^0}, V_{K^0} = V_6 + iV_7$

---

[*] Talk given by S. Oneda

[**] Supported in part by the National Science Foundation under Grant No. NSF GP 8748.

etc), which express[1,3] the usual simple mechanism of SU(3) breaking.

As to the algebra, $[A_i, A_j] = if_{ijk} V_k$, we seek the level schemes of hadrons which realize the asymptotic SU(3) contents of this algebra. For the ground states, the schemes obtained are compatible with those of the SU(6) x O(3) classification.

The sum rules obtained recover most of the _good_ results of SU(6). Neither exact SU(6)[4] nor the saturation of the C.R.'s by low-lying intermediate states is assumed.

We denote the hadrons by $B_{\alpha,s}$ where α stands for the physical SU(3) indices, $(\pi_s, K_s, \eta_s, \eta'_s)$ and $(N_s, \Lambda_s, \Sigma_s, \Xi_s)$, etc., and s for the J^{PC}, etc.

Our asymptotic SU(3) assumes[5] that the annihilation operator $a_{\alpha,s}(\vec{k},\lambda)$ of $B_{\alpha,s}$ (λ denotes helicity) transforms _linearly_ under SU(3) but only in the limit $\vec{k} \to \infty$, i.e.,

$$[V_i, a_{\alpha,s}(\vec{k},\lambda)] = i\Sigma_{\beta,t}\, u_{i\alpha\beta,st}\, a_{\beta,t}(\vec{k},\lambda) + \delta u_{i\alpha}(\vec{k},\lambda),$$

and (1)

$$\delta u_{i\alpha}(\vec{k},\lambda) \to 0 \text{ as } \frac{1}{|\vec{k}|^{1+\epsilon}} \text{ for } \vec{k} \to \infty,\ (\epsilon > 0). \quad (2)$$

In Eq. (1) the summation over t is taken for all the t which have the same J^{PC} or J^P as that of s and the particle mixing in broken SU(3) is taken care of there. Thus at $\vec{k} \to \infty$, $a_{\alpha,s}(\vec{k},\lambda)$ will be related linearly to the usual representation operator $a_{j,s}(\vec{k},\lambda)$ (for the nonet, for example, j = 0, 1, 2,...,8).

Our asymptotic SU(3) is compatible with the imposition of the C.R., $[V_i, V_j] = if_{ijk} V_k$. We summarize our result as follows:

a) $[\dot{V}_{K^0}, V_{K^0}] = 0 \to$ G-M-O mass formula with mixing[1,3]. (3)

The G-M-O formula with "mixing" becomes an exact constraint.

b) $[V_i, A_j] = if_{ijk} A_k \to$ SU(3) behavior of the matrix elements of the axial-vector charge A_i; define $G^i_{ts} = \lim_{\vec{k} \to \infty} <B_t|A_i|B_s(\vec{k})>$, (i = 1, ..., 8). In broken SU(3), G^i_{ts} still allows conventional exact SU(3) plus mixing parametrization[6]. Thus the D/F ratio can still maintain its physical significance for the G^i_{ts}. For the PS-meson coupling sum rules (obtained by using PCAC for the A_i in G^i_{ts}), simple broken SU(3) correction factors now appear and manifest the effect of SU(3) breaking[6]. Example: (i) $(2G_{K^*K^-\pi^0}/G_{\rho\pi^-\pi^0}) = (K^*/\rho)$, (ii) $B(\frac{1}{2}^-) \to B(\frac{1}{2}^+) + \pi$, $(p'-n)^{-1} g_{p'n\pi} = \sqrt{3/2} (\Sigma'-\Lambda)^{-1} g_{\Sigma'\Lambda\pi} - \sqrt{1/2} (\Sigma'-\Sigma)^{-1} g_{\Sigma'\Sigma\pi}$. Notice the mass correction factors.

c) $[\dot{V}_{K^0}, A_{\pi^-}] = [\dot{V}_{K^0}, A_{K^0}] = 0$ etc. $\to$ Inter-SU(3)-multiplet mass relations. Bosons[1]: $K_s^2 - \pi_s^2 = $ const., s is arbitrary, (4)
i.e., $K^2 - \pi^2 = K^{*2} - \rho^2 = K^{**2} - A_2^2 = K_A^2 - A_1^2 = K_B^2 - B^2 = ...$[7].
Baryons[6]: $\Sigma_s^2 - N_s^2 = \Xi_s^2 - \Sigma_s^2 = \delta_t^2 = $ const., (s and t are arbitrary), (5)
where $\Sigma_s^2 = \Lambda_s^2$ and δ_t^2 is the quadratic equal mass spacing of the decuplet. The $\Sigma_s - \Lambda_s$ degeneracy might be removed by introducing the mixing between the low-lying and higher-lying baryons with the same J^P.

d) $[V_{K^0}, A_{\pi^-}] = [\dot{V}_{K^0}, A_{\pi^-}] = 0 \to$ nonet pion coupling sum rules[1].

$$\frac{b}{a} \equiv \frac{<\eta_s|A_{\pi^-}|\pi_u^+(\vec{k})>}{<\eta_s'|A_{\pi^-}|\pi_u^+(\vec{k})>} = \tan\theta_s \cdot \frac{(\eta_s'^2 - \pi_s^2)}{(\eta_s^2 - \pi_s^2)}, \quad \vec{k} \to \infty, \quad (6)$$

s and u are arbitrary ($C_s C_u = 1$) and θ_s is the η_s-η'_s mixing angle. Example: Use PCAC for the A_π- in Eq. (6).

$$\frac{G_{\phi\rho\pi}}{G_{\omega\rho\pi}} = \tan\theta_{\omega\rho} \cdot \frac{(\rho^2-\omega^2)}{(\rho^2-\phi^2)} \simeq 0.053 \rightarrow \Gamma(\phi \rightarrow \rho\pi) \simeq 0.53 \text{ MeV}, \quad (7)$$

$$\frac{G_{f'\pi\pi}}{G_{f\pi\pi}} = \tan\theta_{ff'} \cdot (\frac{f'^2}{f^2}) \frac{(f^2-\pi^2)}{(f'^2-\pi^2)} \frac{(f^2-A_2^2)}{(f'^2-A_2^2)} \simeq 0.17 \rightarrow \frac{\Gamma(f' \rightarrow \pi\pi)}{\Gamma(f \rightarrow \pi\pi)} \simeq 5.4\%. \quad (8)$$

The interesting interplays, $\rho \simeq \omega \leftrightarrow G_{\phi\rho\pi} \simeq 0$ and $f \simeq A_2 \leftrightarrow G_{f'\pi\pi} \simeq 0$ are obtained and, moreover, allow numerical predictions.

e) <u>Asymptotic Algebraic</u> realization of SU(3) in the $[A_i, A_j] = if_{ijk} V_k$ and the level scheme of hadrons. Insert this C.R. between $\langle B_{\alpha,s}(\vec{p})|$ and $|B_{\beta,s}(\vec{p})\rangle$ with $\vec{p} \rightarrow \infty$. The r.h.s. of this equation produces definite pure numbers $g_{\alpha\beta}$ according to our <u>asymptotic</u> SU(3). If one varies α and β, the ratios of these (non-vanishing) $g_{\alpha\beta}$'s express the asymptotic SU(3) contents of the algebra. Write the l.h.s. as the sum over the intermediate states inserted between the A_i and A_j. We assume that the hadron states make a certain grouping, $R_0, R_1, \ldots$, and in the intermediate states each R_i <u>separately</u> realizes the ratios of the $g_{\alpha\beta}$'s.

Bosons[8]: "Each boson forms a nonet ($\pi_s, K_s, \eta_s, \eta'_s$) and <u>separately</u> realizes the <u>asymptotic</u> SU(3) contents of the algebra." (This is expressed in Eq. (12).) We summarize the sum rules for the bosons as follows,

$$\sin^2\theta_s = \frac{(3\eta_s^2 - 4K_s^2 + \pi_s^2)}{3(\eta_s^2 - \eta_s'^2)} \leftarrow [\dot{V}_{K^0}, V_{K^0}] = 0 \text{ (see Eq. (3))}, \quad (9)$$

$$\frac{\sqrt{3}}{2}\sin\theta_s \cdot a - \frac{\sqrt{3}}{2}\cos\theta_s \cdot b = -c \leftarrow [V_{K^0}, A_{\pi^-}] = 0, \quad (10)$$

$$\frac{\sqrt{3}}{2}\sin\theta_s \cdot (K_s^2 - \eta_s'^2) a - \frac{\sqrt{3}}{2}\cos\theta_s \cdot (K_s^2 - \eta_s^2) b = (K_s^2 - \pi_s^2)c$$

$$\leftarrow [V_{K^0}, A_{\pi^-}] = 0, \quad (11)$$

$$\frac{1}{2}(a^+ a + b^+ b) = c^+ c, \text{ (spin sum is understood)}$$
$$\leftarrow [A_i, A_j] = if_{ijk}V_k. \quad (12)$$

Here $a \equiv \langle \eta_s'(\vec{p})|A_{\pi^-}|\pi_u^+\rangle$, $b \equiv \langle \eta_s(\vec{p})|A_{\pi^-}|\pi_u^+\rangle$ and $c \equiv \langle K_s^0(\vec{p})|A_{\pi^-}|K_u^+\rangle$ with $\vec{p} \to \infty$ and $C_s C_u = 1$. The remarkable property of this set of sum rules is that they are consistent with the so-called <u>ideal nonet</u> scheme. In our notation, the scheme implies, (i) $\eta_s'^2 = \pi_s^2$, (ii) $\eta_s^2 - K_s^2 = K_s^2 - \pi_s^2$, (iii) $\sin^2\theta_s = 1/3$ (canonical mixing angle), (iv) $b \equiv \langle \eta_s'(\vec{p})|A_{\pi^-}|\pi_u^+\rangle = 0$, $\vec{p} \to \infty$. If one of these constraints of ideal nonet (such as $\rho \sim \omega$ or $G_{f'\pi\pi} \simeq 0$ etc.) turns our to be satisfied, then Eqs. (9) - (12) yields the <u>other three</u> ideal nonet constraints <u>automatically</u>. Therefore, the sum rules Eqs. (9) - (12) are consistent with our knowledge about the well-established 1^{--} and 2^{++} mesons which constitute an (approximate) <u>ideal</u> nonet. However, Eqs. (9) - (12) do not require that every boson nonet follow the ideal nonet pattern. They also permit the weakly coupled nonet structure. Eqs. (9) - (12) predict the following nonet mass relation

$$\eta_s'^2 - \pi_s^2 = \frac{-4(K_s^2 - \pi_s^2)(2K_s^2 - \eta_s^2 - \pi_s^2)}{\pi_s^2 + 3\eta_s^2 - 4K_s^2}, \quad (13)$$

which turns out to coincide with the Schwinger's nonet formula[9]. For the PS-meson mass Eq. (13) predicts the ninth PS-meson closer to the E(1422) (which can also be

0^{-+}) than to the X(958).

Baryons[10]: the simplest choice of the ground state R_o is $R_o(\frac{1}{2}^+, \frac{3}{2}^+)$. The asymptotic algebraic realization of SU(3) requires for the part coming from the R_o;

$$2G^2(D+F)^2 - 2G^{*2} = \frac{2}{3}G^2 D^2 + 2G^2 F^2 + \frac{1}{4}G^{*2} = 2G^2(D-F)^2 + G^{*2} = f_o, \qquad (14)$$

where $<p|A_{\pi^+}|n(\vec{p})> = G\sqrt{2}\,(D+F) \equiv g_A(0)$, $<\Sigma^+|A_{\pi^+}|\Lambda^o(\vec{p})> = G(2/\sqrt{3})D$, $<\Sigma^+|A_{\pi^+}|\Sigma^o(\vec{p})> = G(-2F)$ etc. with $\vec{p} \to \infty$ and $D + F = 1$. (See the discussion in (b)). G^* is defined by $<p|A_{\pi^+}|\Delta^o><\Delta^o|A_{\pi^-}|p(\vec{p})> = G^{*2}$, $\vec{p} \to \infty$. f_o is the <u>universal</u> fractional contribution from the R_o to the sum rules. We then obtain,

$$\frac{D}{F} = \frac{3}{2}, \quad |G^*| = |\frac{4}{5}G| = |\frac{4}{5\sqrt{2}} g_A(0)|, \qquad (15)$$

$$g_A(0) = \frac{5}{3}\sqrt{f_o}\,. \qquad (16)$$

Thus assuming <u>neither</u> exact $SU(6)^4$ <u>nor</u> the <u>complete</u> saturation[11] of the C.R. only by the members of R_o, we have derived the good result of SU(6), Eq. (15). These assumptions previously made[4,11] are certainly inaccurate as exemplified by their apparent bad result $g_A(0) = 5/3$. On the contrary, the value $f_o \sim 50\%$ yields $g_A(0) \sim 1.2$ in Eq. (16). The value of G^* corresponding to $f_o \sim 50\%$ (Eq. (15)) predicts a reasonable value $\Gamma(\Delta \to N\pi) \sim 150$ MeV, if PCAC is used. Thus as far as the ground state R_o is concerned, everything is tied together neatly. The level scheme of higher excited states remains to be investigated.

We thank Professor H. Umezawa and our colleagues at the University of Maryland for many useful discussions.

REFERENCES

1. S. Oneda and Seisaku Matsuda, Nucl. Phys. B26 (1971) and the earlier references cited there.
2. M. Gell-Mann, Physics 1, 63 (1964).
3. S. Fubini and G. Furlan, Physics 1, 229 (1965); G. Furlan, F. Lannoy, G. Rossetti and G. Segre, Nuovo Cimento 40, 597 (1965); K. Nishijima and J. Swank, Nucl. Phys. B3, 553 (1967).
4. F. Gursey and L. A. Radicati, Phys. Rev. Letters 13, 173 (1964); B. Sakita, Phys. Rev. 136, B 1756 (1964).
5. S. Oneda, H. Umezawa and S. Matsuda, Phys. Rev. Letters 25, 71 (1970).
6. S. Oneda and Seisaku Matsuda, Phys. Rev. D2, 887 (1970). See also G. Fourez, Nucl. Phys. B18, 189 (1970).
7. We have presented an argument that Eq. (14) is also valid for the 1^{++} - meson. T. Laankan and S. Oneda, University of Maryland, Technical Report No. 73-038, 1972, to be published.
8. S. Oneda and Seisaku Matsuda, Phys. Letters 37B, 105 (1971).
9. J. Schwinger, Phys. Rev. Letters 12, 237 (1964). See also, Riazuddin and K. T. Mahanthappa, Phys. Rev. 147, 972 (1966). D. Horn, J. J. Coyne, S. Meshkov and J. C. Carter, ibid., 147, 980 (1966); A. N. Zaslavsky, V. I. Ogievetsky and V. Tybor, JETP Letters 6, 106 (1967); V. I. Ogievetsky, Phys. Letters 33B, 227 (1970).
10. S. Oneda and Seisaku Matsuda, Phys. Rev. D5, 2287 (1972).
11. I. S. Gerstein, Phys. Rev. Letters 16, 114 (1966).

THE ELECTRON AND THE MUON AS EIGENSOLUTIONS IN QUANTUM ELECTROMECHANICS

Fritz Bopp and W. Lutzenberger

Sektion Physik der Ludwig-Maximilians-Universität

München (presented by Fritz Bopp)

The classical motion of a charged mass point in its own electromagnetic field may be described by two different, however mathematically equivalent, systems of equations: (i) by the usual equations of motion and the field equations, (ii) by equations of motion with retarded forces, obtained by eliminating the field. Both systems of equations may be quantized. We ask the question: What happens, if we quantize the second system? In particular, we consider the motion of one charged mass point in its own field without radiation damping.

Replacing the Green's function of Maxwell's theory $\delta(x^2)$ [1] by $\delta(x^2 + \ell^2)$ [2], where ℓ is a certain length constant, we obtain as interaction contribution to the Lagrangian[3]:

$$L = -\frac{\alpha}{2} \frac{\dot{x}^2 + \dot{x} \cdot \dot{y}}{\dot{x} \cdot y} \quad , \quad y^2 = -\ell^2 \quad , \tag{1}$$

(units: $\hbar$, c; signature: -+++, α = fine structure constant). This means: (i) the Lagrangian is finite even in the limit $\ell \to 0$, and (ii) the number of degrees of freedom is twice that of a Newtonian particle, and not infinite. Both results occur without any approximation. Both would be hidden if we were to start with a retardation development.

Adjoining the auxiliary condition $\dot{y} \cdot y = 0$ by a Lagrangian parameter, eq. (1) may be replaced by

$$L = - \frac{\alpha}{2} \frac{\dot{x}^2 + \dot{x} \cdot \dot{y}}{\dot{x} \cdot y} + \lambda \, \dot{y} \cdot y \quad . \tag{2}$$

Hence, we get the relativistic Hamiltonian

$$H = p \cdot v - v^2 = 0 \, , \, v = q - \frac{1}{y^2} (q \cdot y + \frac{\alpha}{2}) \, , \tag{3}$$

where (p, x) and (q, y) are the canonically conjugate coordinates. The constant value of the Lorentz invariant Hamiltonian must be equal zero, because the interaction integral $\int L d\tau$ is parameter invariant.

We consider only the case $y^2 = -\ell^2$, $\ell = 1$, taking ℓ as unit of length. Inserting the canonical transformation

$$x \to x - \tfrac{1}{2} y \, , \, y \to y \, , \, p \to p \, , \, q \to q + \tfrac{1}{2} p$$

and introducing c.m.s. values

$$p^0 = m \, , \, \underline{p} = 0 \, ,$$

we obtain as expression for the mass:

$$m = \frac{1}{\underline{y}^2} \sqrt{\alpha^2 + 4 \underline{y}^2 \underline{Q}^2} + \frac{\alpha}{\underline{y}^2} \sqrt{1 + \underline{y}^2} \, ,$$

$$\underline{Q} = \underline{q} + \frac{1}{\underline{y}^2} (\sqrt{1 + \underline{y}^2} - 1)(\underline{q} \cdot \underline{y}) \underline{y} \quad . \tag{4}$$

Using Dirac's method of quantizing momenta square roots, the following wave equation appears:

$$\{\frac{\alpha}{y^2}\rho_3 + 2\rho_2\underline{\sigma}\cdot:(\frac{1}{y}\underline{Q}^2): + \frac{\alpha}{y^2}\sqrt{1+y^2}+M\rho_3\}\psi(\underline{y})=m\psi(\underline{y}), \quad (5)$$

in which we have introduced the bare mass M ad hoc. (Clearly, that must be done later in a more fundamental way.) Subsequently we assume that $M = \pm|M|$, where $|M|$ should be a fundamental constant.

The integration of eq. (5) is straightforward. The angles of the vector $\underline{y}$ can be separated as in the relativistic H-atom. The quantum numbers $k = \pm 1$, $\pm 2, \ldots$ occur, belonging to the states $S_{1/2}$, $P_{1/2}$, $P_{3/2}$, $D_{3/2} \ldots$, and we obtain finally the following two-component equation

$$\frac{du}{dt} = \frac{1}{2}(\frac{t+1}{t-1})^k (M+m-\frac{\alpha}{t+1}) v ,$$

$$\frac{dv}{dt} = \frac{1}{2}(\frac{t-1}{t+1})^k (M-m+\frac{\alpha}{t-1}) u , \quad (6)$$

$$t = \sqrt{1+y^2}$$

which may be solved by numerical integration.

To each spectral term $S_{1/2}$, $P_{1/2}$, $P_{3/2} \ldots$, and to each sign of the bare mass $M = \pm|M|$ belongs a mass spectrum. All excited state masses nearly equal the bare mass. However the ground states may be shifted appreciably. If we take into account the selection rules for dipole radiation, only two of these states are electromagnetically stable:

$$\begin{aligned} M &= -|M| , & k &= +1 , & S_{1/2} , \\ M &= +|M| , & k &= -1 , & P_{1/2} . \end{aligned} \quad (7)$$

Therefore, we may guess that these two states are those of the electron and the muon. Hence numerical integration yields

$$M = 7.90 \times 10^{-90} = 15.32\ m_p,$$
$$\ell = 0.489 \times 10^{-90}\ m_p^{-1}, \qquad (8)$$

and we obtain the following mass table of the ground states:

^2S_j	^2P_j	^2D_j	^2F_j
m = 1	206.3	305.3	377.1 (m_e units)

($j = 1 \mp \tfrac{1}{2}$, $k = \{{j+1/2 \atop -j+1/2}$, $M = \mp |M|$). The unknown particles with muonic masses may have dipole lifetimes of about 10^{-18} to 10^{-17} sec. and are not easily observed. The excited states with lifetimes of about $10^{-27} - 10^{-26}$ sec. are still unobservable. In this case the line width exceeds the mass values. Observing the $^2P_{3/2}$-state with $m \simeq 206\ m_e$ would be a test of our hypothesis.

Many questions remain open. Some assumptions require further considerations ($\ell \neq 0$, $M = \pm |M|$, the e-μ-guess). However, the essential results are well settled: no singularities even in the limit $\ell \to 0$, a finite number of degrees of freedom, and exact solutions by avoiding perturbation theory.

REFERENCES

1. P.A.M. Dirac, Proc. Roy. Soc. (Ld.) $\underline{A}$ $\underline{197}$, 148 (1938).
2. E. Groschwitz, Z. Naturf. $\underline{7a}$, 658 (1952), Eq. (3.4).
3. A.D. Fokker, Z. Phys. $\underline{58}$, 386 (1929).

SLOW POSITRON COLLISION IN GASES

H. S. W. Massey

Department of Physics and Astronomy

University College London

The study of the rates of the various collision processes associated with positrons with energy ranging from thermal up to a few tens of eV is of interest, not only for its own sake but because of the more stringent tests which it imposes on theoretical approximations than does the corresponding study of electron collisions. For example, the elastic scattering of slow electrons by atoms is determined largely by the combined effect of (a) the mean static field of the unperturbed atom (b) the dipole distortion of this field by the incoming electron and (c) electron exchange. Of these the first two always give rise to attractive forces. On the other hand for positrons (a) is always repulsive and is opposed by the dipole distortion (b) which is attractive as for electrons. The accurate evaluation of the net effect of (a) and (b) is often more difficult when they act in opposite sense. For positron collisions, exchange is replaced by virtual positronium formation which tends to provide additional attraction but is difficult to calculate accurately because it depends strongly on electron-positron correlation.

In addition it is possible to observe certain quantities for positron collisions which have no analogue for electrons, depending as they do on positron annihilation rates. These again depend very strongly on electron-positron correlation and so demand an accurate theory for their description and prediction.

Finally, the chemistry of positronium is of special interest as it contains in the extreme the features which are only barely becoming apparent with hydrogen - for example the unusual features associated with the hydrogen bond.

This account will confine itself to some of the less complicated theoretical aspects of the subject but will deal with some quite sophisticated experimental techniques which have recently made it possible to derive more directly interpretable physical data than hitherto. The subject is unusual in that, while the theoretical physics involved is that of atomic collisions, the experimental techniques are essentially those of nuclear physics in so far as the central observational method is the use of sophisticated particle counter techniques.

<u>Observable quantities</u>. The rate of annihilation of positrons in an assembly of electrons is given by

$$\lambda = 4\pi r_o^2 c n_s ,$$

where n_s is the average number of electrons per unit volume at the position of the positron possessing opposite spin to it. $r_o (=e^2/mc^2)$ is the classical radius of the electron and c is the velocity of light. If n is the concentration of atoms per unit volume we may write

$$n_s = Z_a n$$

where Z_a, the effective number of annihilation electrons per atom, will in general be a function of the relative velocity v of the positrons and atoms. We may now define an effective annihilation cross section

$$Q_a = \pi r_o^2 c Z_a / v$$

Direct measurements may be made of the mean value of Z_a averaged over a positron velocity distribution, as a function of the R.M.S. velocity.

A further quantity associated with annihilation which may be observed is the directional correlation between the two gamma quanta which result. This depends on the distribution of relative velocity between the positrons and the electrons with which they annihilate.

Until recently it has not been possible to measure directly the total cross section for collisions of positrons in gases but new methods, which will be described later, now make this possible for positrons with velocities which are prescribed within limits of the order of 1 eV.

Indirect information is also available about the momentum transfer or diffusion cross section, as is explained below.

<u>Mathematical formulae for cross sections</u>. Consider the collisions of positrons with atoms containing Z electrons whose coordinates relative to the nucleus are $\underset{\sim}{r}_1, \ldots \underset{\sim}{r}_Z$ $\underset{\sim}{r}$ being that of the positron. For simplicity we restrict ourselves to positron energies which are so low that only elastic collisions are possible. The overall wave function for the system will have the asymptotic form

for large r

$$\Psi(r_1 \ldots r_Z; r) \sim (e^{ikr} + f(\theta) r^{-1} e^{ikr}) \psi(r_1, \ldots r_Z)$$

where ψ is the wave function for the ground state of an atom and k is the wave number of the positron motions.

The total elastic cross section is then given by

$$Q_t = 2\pi \int_0^\pi |f(\theta)|^2 \sin\theta \, d\theta \quad,$$

the momentum transfer or diffusion cross section by

$$Q_d = 2\pi \int_0^\pi (1 - \cos\theta) |f(\theta)|^2 \sin\theta \, d\theta \quad,$$

and the effective number of annihilation electrons per atom by

$$Z_a = \int |\Psi|^2 \sum_{i=1}^{Z} \delta(r_i - r) dr_1 \ldots dr_Z \, dr$$

If $P(q_z)$ is the distribution of a single component of momentum derived from angular correlation observations of the two gamma decay

$$P(q_z) = \iint |S(q)|^2 \, dq_x \, dq_y$$

where

$$S(q) = \int \sum_i S(r_i - r) e^{i q \cdot r} |\Psi|^2 \, dr_1 \ldots dr_Z \, dr$$

The importance of correlation effects in determining Z_a and $P(q_z)$ is manifest from these formulae. Thus Z_a and $P(q_z)$ depend more strongly on the form assumed for Ψ than do Q_t and Q_d.

As an example, for positron – H atom collisions in the low velocity limit Ψ is usually represented approximately in the form of a variational wave function Ψ_t of the Schwartz type

$$\Psi_t = \sum_{lmn} c_{lmn}\, r^l r_1^m\, |r-r_1|^n\, e^{-\alpha(r+r_1)} + \text{polarization terms}.$$

In Table 1 the convergence of the calculated scattering length $a = (Q_t/\pi)^{1/2}$ and of Z_a with inclusion of increasing number of linear coefficients is compared.[1]

Table 1

Number of linear coefficients c_{lmn}	4	10	20	30
Scattering length (a_o)	-0.68	-1.73	-2.02	-2.09
Z_a		1.37	5.15	7.51	8.48

Further calculations with more elaborate functions suggest that Z_a converges to about 8.9. It will be seen that even with trial functions containing correlation terms the convergence for Z_a is considerably slower than for the scattering length. It is also of interest to notice that Z_a is nearly an order of magnitude larger than the actual number of atomic electrons. This is due to the importance of dipole polarization of the atom by the positron which increases quite strongly the overlap with the electron charge density.

<u>The slowing down of positrons in gases – positronium formation</u>. Positrons emitted from a radioactive source with energy of a hundred keV or so will initially slow down in passage through a gas in much the same way as other charged particles, through energy loss by inelastic

collisions with the gas atoms. The possibility of positronium formation by capture of electrons from atoms on impact is not important at first. Thus the threshold energy for positronium formation is $E_i - E_{ps}$ where E_i is the ionization energy of an atom and E_{ps} the binding energy of positronium (6.8 eV). If the positron energy before capture exceeds E_i the positronium produced will have a kinetic energy greater than E_{ps} and so can be broken up again in further collisions with gas atoms. On this basis, "permanent" positronium production will mainly occur through capture of positrons with energy E such that

$$E_{ex} > E > E_i - E_{ps}$$

The range between E_{ex} and $E_i - E_{ps}$ is known as the Ore gap. Positrons passing through the gap without capturing an electron can only slow down through momentum transfer in elastic collisions until they come to thermal equilibrium with the gas atoms.

Until recently almost all experimental work on positrons in gases was concerned with studying these final stages in the life of a positron. In the last six months or so it has proved possible to measure total cross sections for collision of positrons of well defined energy with gas atoms, a new development of considerable interest and importance.

Experimental detection of positrons and positron annihilation. The usual source of positrons used for this work is β^+ radioactive Na^{22}. The beginning and end of the life of a positron has usually been recorded by observing electronically the delayed coincidence between production of a 1.28 MeV γ-ray which is emitted promptly after the β^+ decay and of a γ-ray emitted as the positron undergoes an annihilation collision with an electron. Recently the positron research group at University College London comprising T. C. Griffith, G. R. Heyland, K. F. Canter and P. G. Coleman[2] have improved this technique primarily by replacing the 'start' signal from the prompt γ-ray by one due directly to passage of the positron from the source to the experimental chamber through a thin scintallator. This increases the chance of timing a positron by an order of magnitude.

As a result of the fast counting rate of the positron counter the 'start' signal in real time is delayed for an interval longer than the time range of interest and used as a 'stop' in the timing sequence. This leads to an inverted time spectrum and the prompt events, normally not of much interest in the analysis, can be vetoed, thus minimizing the generation of random background coincidences. In the data analysis careful consideration has been given to correct evaluation and subtraction of the remaining random coincidence background. These improvements in technique not only make possible much more accurate measurement of such parameters as decay rates but also the observation of other features which would otherwise be lost by the limited statistical accuracy of the data.

The measurement of total cross sections for collision of positrons with gas atoms. The vital component of an experiment to measure total cross sections for positron collisions as a function of positron velocity is a source of positrons with controllable and definable energy of a few eV which is intense enough to be used for transmission experiments in gases.

It can hardly be said that the development of such

a source has followed an orderly and logical process. Indeed it is still far from clear how the present source, which has proved adequate for the purpose, works at all. Its origin dates back from experiments by McGowan and his colleagues directed towards the energy moderation of positrons produced through pair production from γ-rays produced by electrons from a linear accelerator. They obtained evidence[3] that if the positrons were slowed down by passage through a mica sheet coated with gold, there emerged on the far side slow positrons in a quite narrow range about a maximum value of a few eV. It was suggested that these positrons arose by expulsion through the gold surface of thermalized positrons accelerated outwards by the work function.

Efforts to reproduce this effect at University College, London, with positrons from a Na^{22} source initially proved unsuccessful but it was found that if the positrons emerging from a thin scintillator were then allowed to suffer back scattering from a gold surface, slow positrons emerged in useful numbers with quite narrow energy distribution about a few eV maximum.

Advantage was taken of this to design a suitable source for transmission experiments.[4] The general arrangement is shown in Fig. 1. Positrons from the source pass through a plastic scintillator of 0.17 mm thickness and a Melinex window of 0.075 mm thickness, aluminized on the surface in contact with the scintillator, and an aluminum foil of thickness 0.0008 mm, to enter a copper cylinder 20 mm long coated internally with gold to a thickness of about 0.025 mm. At the far end of the cylinder is a fine grid followed by an insulating ring and a second grid at earth potential. By applying a positive potential to the moderator, posi-

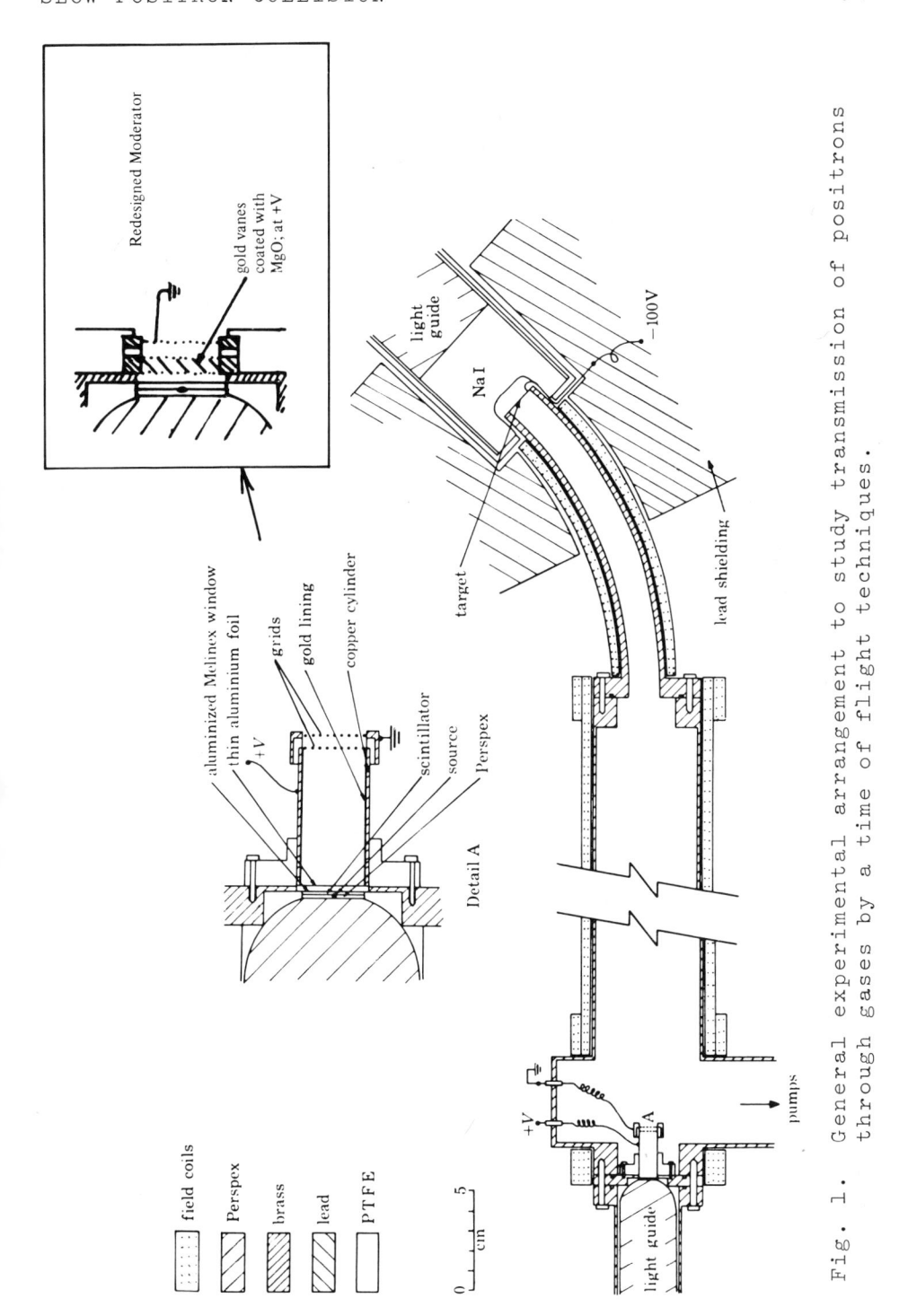

SLOW POSITRON COLLISION 197

Fig. 1. General experimental arrangement to study transmission of positrons through gases by a time of flight techniques.

trons emerging (from the gold) and reaching the far end of the cylinder are accelerated to the desired energy.

The accelerated positrons enter the time of flight tube which includes a straight section of 700 mm length followed by a section of 150 mm length curved in an arc of 250 mm radius. By a suitable system of coils, magnetic fields are provided which confine the positrons to helical paths close to the axis throughout the entire flight path. They are detected by annihilation in an aluminum foil enclosed in a sodium iodide counter. The final curved section of the flight tube ensures that the detector is not exposed directly to the source.

It was found that performance of the source could be improved by evaporating magnesium oxide on to the gold surface. In later experiments (see inset Fig. 1) the cylinder was replaced by a series of vanes as in a multiplier of venetian blind type. Approximately 1 in 10^5 of the positrons emitted from the source entered the flight tube as slow positrons.

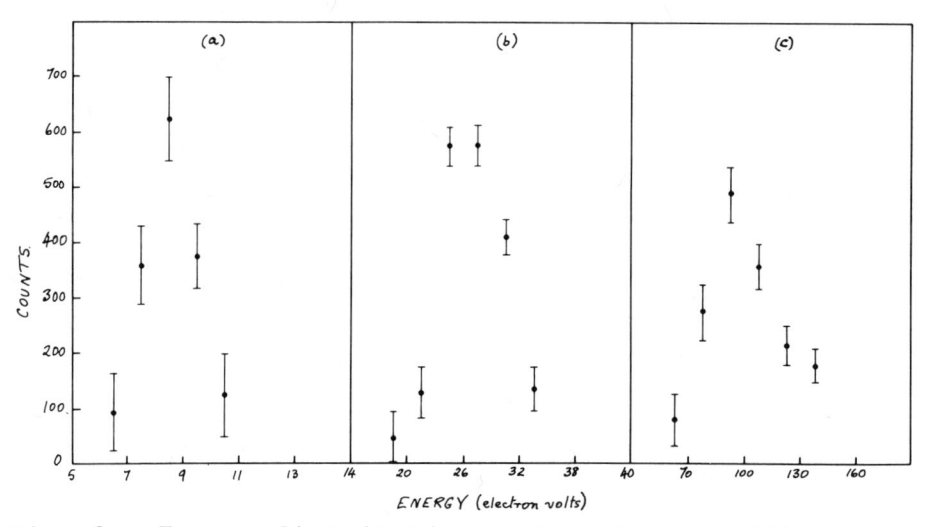

Fig. 2. Energy distributions of positrons with accelerating potentials of (a) 8V (b) 30V and (c) 80V on the gold moderator.

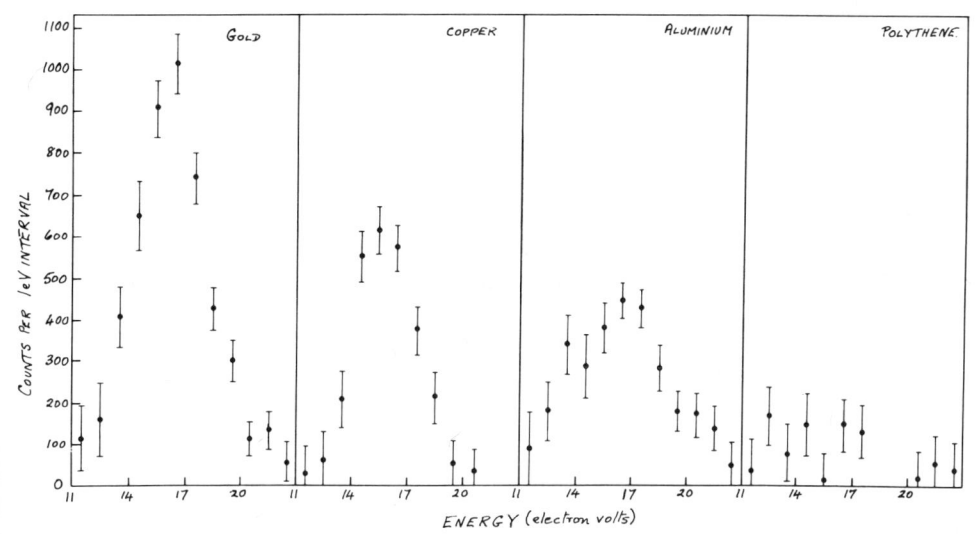

Fig. 3. Energy distribution and yield of positrons from sources in which the inner, back-scattering, surface of the cylinder is coated with different materials.

Fig. 2. shows a typical energy distribution of positrons emerging from the source as determined from the delayed coincidence spectrum observed with the apparatus as in Fig 1. Fig. 3 compares the distributions obtained with a gold coating with those observed with this coating replaced by copper, aluminum and polythene respectively. It will be seen that, of these, gold gives the narrowest distribution and also the largest positron flux.

The distributions shown in Figs. 2 and 3 were obtained with a high vacuum maintained in the flight tube. Fig. 4 shows the effect of introducing helium at a pressure of 2.1×10^{-3} torr into the tube which is apparent in a reduction of the transmission. From analysis of data of this kind the total cross section may be determined as a function of positron energy.

Results obtained in this way for helium[5] are shown

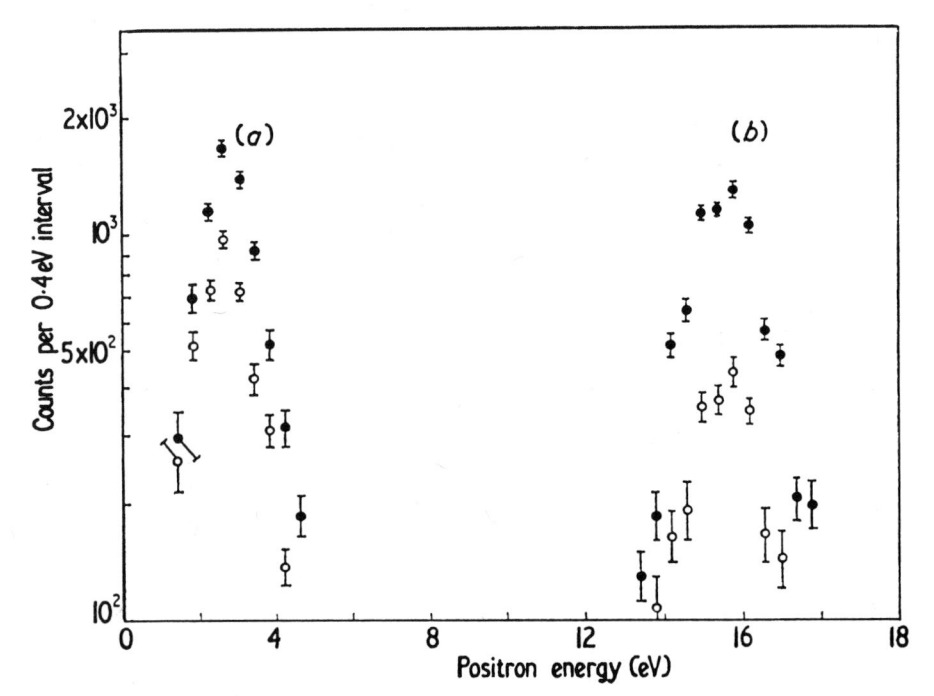

Fig. 4. Energy distribution of positrons with accelerating potentials of (a) 2V (b) 15V in the vanes (see inset Fig. 1) ● at a pressure of 3×10^{-5} torr ○ at a pressure of 2.1×10^{-3} torr of helium.

in Fig. 5 and on a logarithmic scale in Fig. 6 together with results for neon, argon and krypton. It will be seen that the cross sections behave in a rather similar manner at low energies and may well exhibit a Ramsauer minimum at energies below 2 eV which are not accessible to the present technique. Theoretical evaluation of at least the helium cross section should be practicable, and it is already clear that the general shape of the observed cross section is in agreement with available theory. For all the rare gases the new data should help considerably towards a detailed and definite analysis of data obtained from analysis of annihilation spectra which we now describe.

<u>Annihilation spectrum</u>. A major source of information

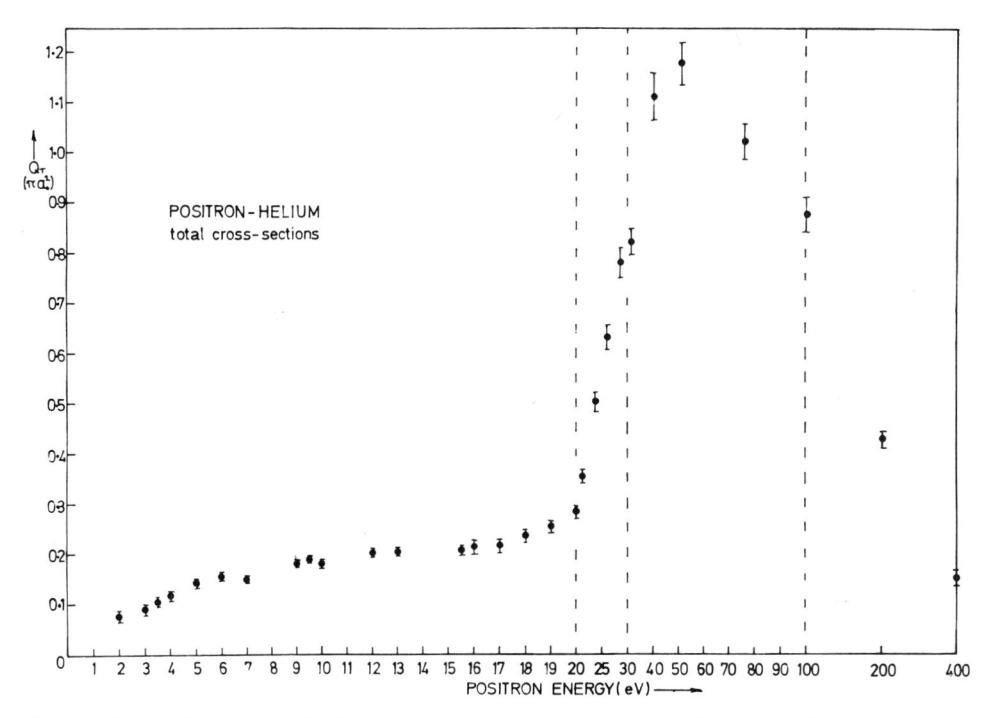

Fig. 5. Observed total cross sections for positron collisions in helium.

about the rates of positron reactions is the annihilation spectrum of positrons in gases at comparatively high pressures so that there is a high probability that positrons will come to thermal equilibrium in the gas before annihilation. Essentially what is measured is the rate of annihilation as a function of time from birth for a positron. Fig. 7 shows a typical set of observations in helium at $300^{\circ}K$ at a concentration of 3.47×10^{21} cm^{-3}. These consist of the number of counts recorded in a given time, within different channels each of which corresponds to a definite interval since the start signal recording the birth of a positron.

The spectrum consists of a prompt peak due to positrons annihilating in the solid material over the source, to parapositronium which has a lifetime of only

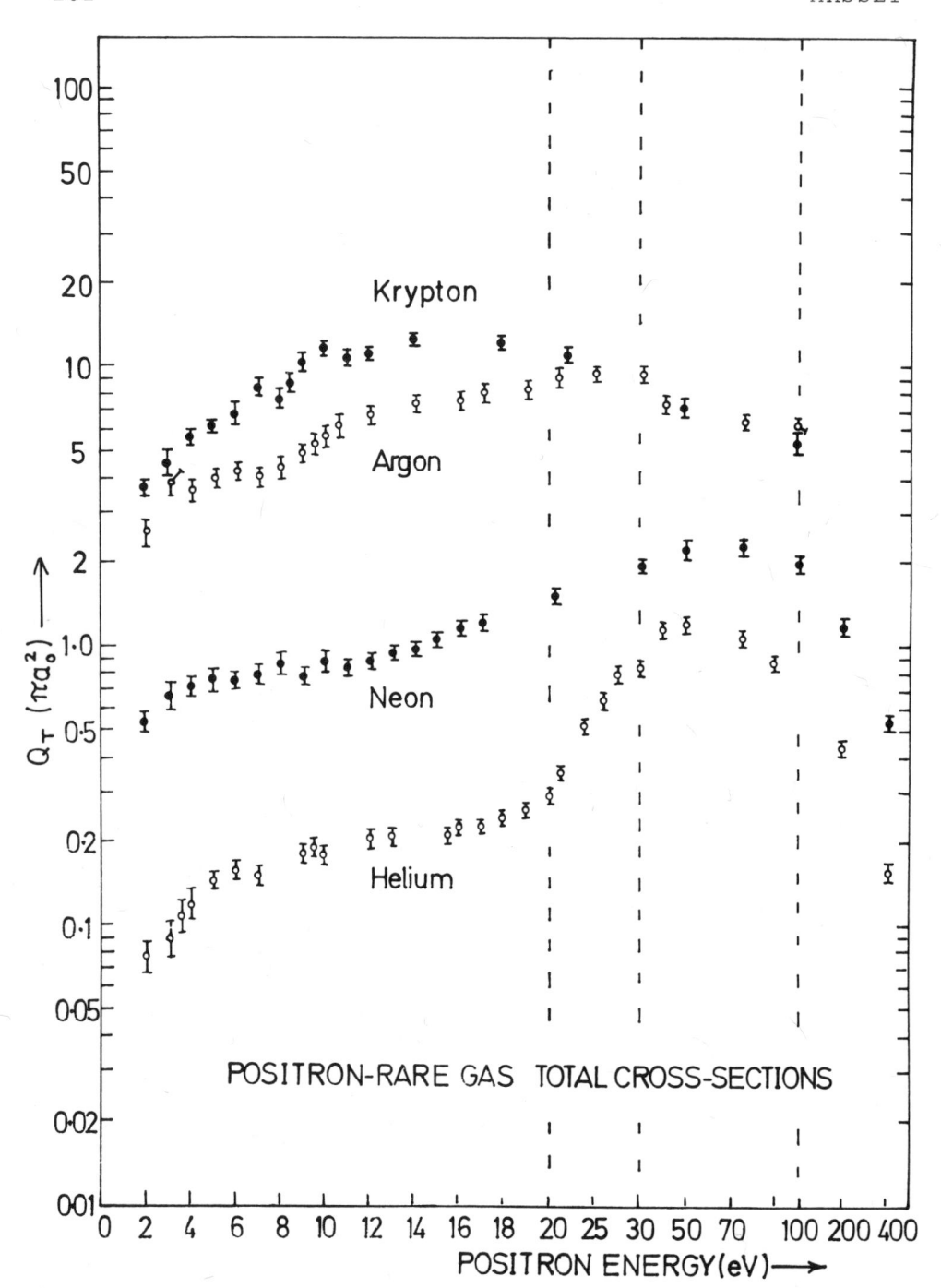

Fig. 6. Observed total cross sections for positron collisions in He, Ne, Ar and Kr.

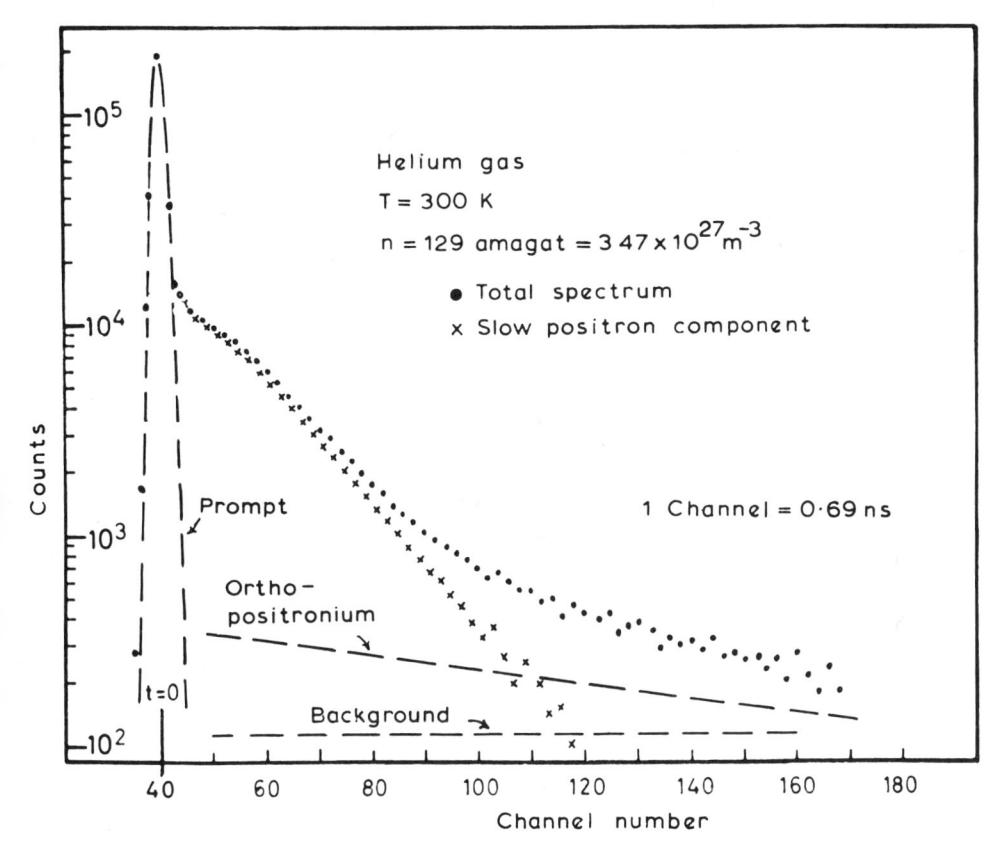

Fig. 7. Typical annihilation spectrum for positrons in helium at room temperature.

10^{-10} sec, and to other background effects. Immediately beyond the prompt peak is the so called shoulder region which extends over the time interval in which the positrons are slowing down to thermal energies. After this stage, annihilation events arise from annihilation of thermalized positrons at a rate λ_a^f and of orthopositronium at a rate λ_a^o. The rate of annihilation thus falls off as

$$a \exp(-\lambda_a^f t) + b \exp(-\lambda_a^o t) \quad ,$$

where a and b refer to the free positron and orthoposi-

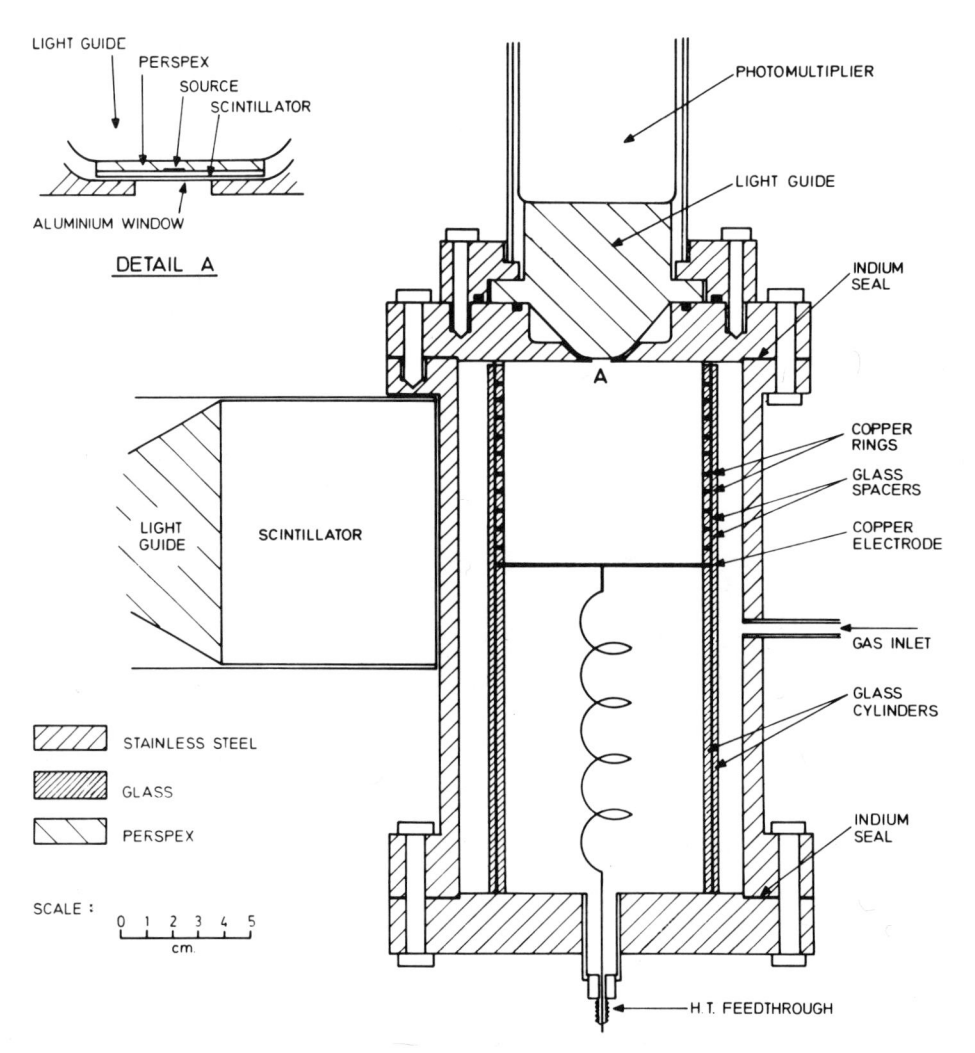

Fig. 8. General experimental arrangement of apparatus used at University College, London, to study positron annihilation spectra.

tronium concentrations at time t = 0. To separate these two terms use is made of the fact that $\lambda_a^o < \lambda_a^f$ so that ultimately the count rate falls off as $\exp(-\lambda_a^o t)$. Nevertheless the separation is quite difficult to carry out unambiguously and accurately and it calls for precise annihilation spectra out to 10^{-7} or larger. Fig. 7 shows the resolution into the thermalized free positron decay, the ortho-positronium decay and the background signal. It is clear, however, that data at later times are less accurate than desired.

Spectra of this kind may be taken in the presence of an electric field E in which case the equilibrium energy of the free positrons and the shape of their velocity distribution will be a function of E/n where n is the atom concentration.

The group at University College, London, have improved the accuracy of measurements of this type by introducing the new start signal technique referred to earlier. Fig. 8 shows the general arrangement which is being used. The annihilation radiation is detected by the large scintillator at the side of the chamber containing the gas under study.

Fig. 9 shows a spectrum obtained in krypton at a preliminary stage with this equipment. The shoulder region is clearly shown. Fig. 10 shows the free positron equilibrium decay rate λ_a^f derived from analysis of annihilation spectra, as a function of gas density for Ar, Kr and N_2. The observed data lie very well on straight lines passing through the origin, as they should. From such plots the mean effective number $\bar{Z}_a$ of annihilation electrons per atom for thermalized positrons may be obtained. For example, for argon at room temperature $\bar{Z}_a = 27.3 \pm 0.2$.

Fig. 11 shows similar plots of λ_a^o for Arcton 12, Freon 12 and argon. In these cases the straight lines extrapolate at zero gas density to a value $\lambda_a^o(0)$ which is the natural decay rate of orthopositronium. From the results shown in Fig. 11 $\lambda_a^o(0) = (7.262 \pm 0.015)\mu sec^{-1}$

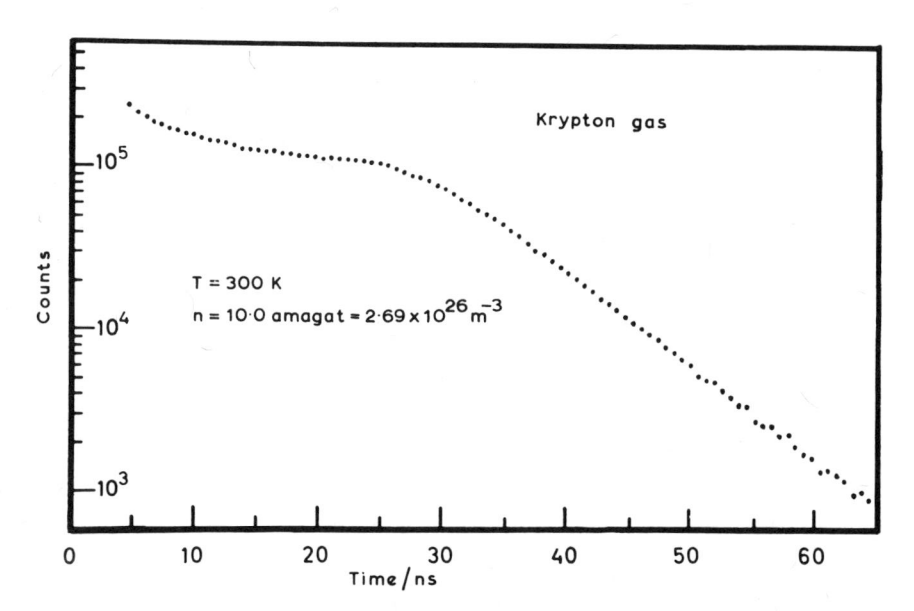

Fig. 9. Annihilation spectrum in krypton observed at room temperature with the apparatus shown in Fig. 8.

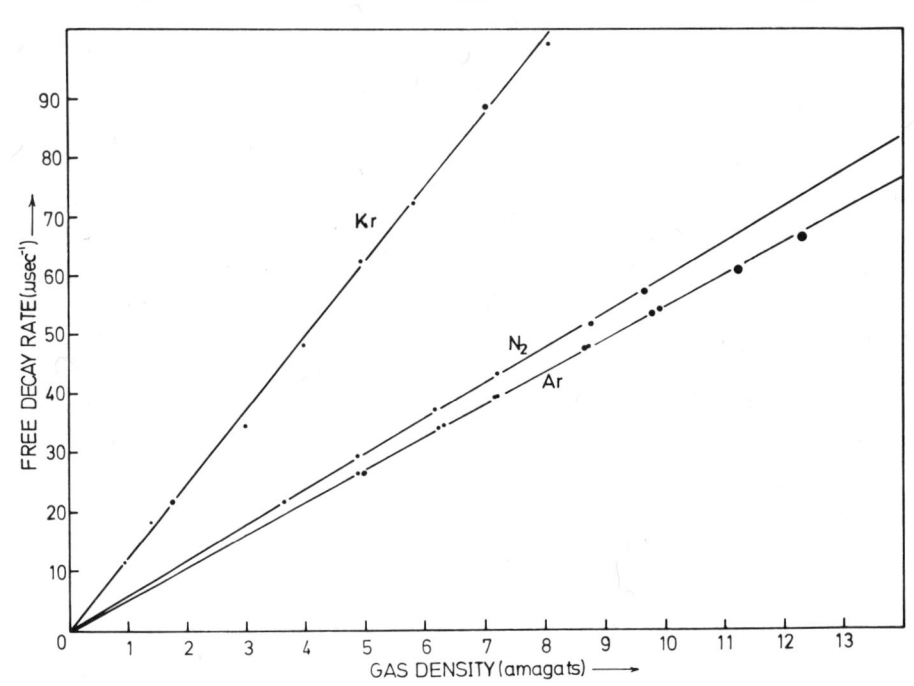

Fig. 10. Observed annihilation rate of free positrons in Ar, Kr and N_2 as a function of gas density.

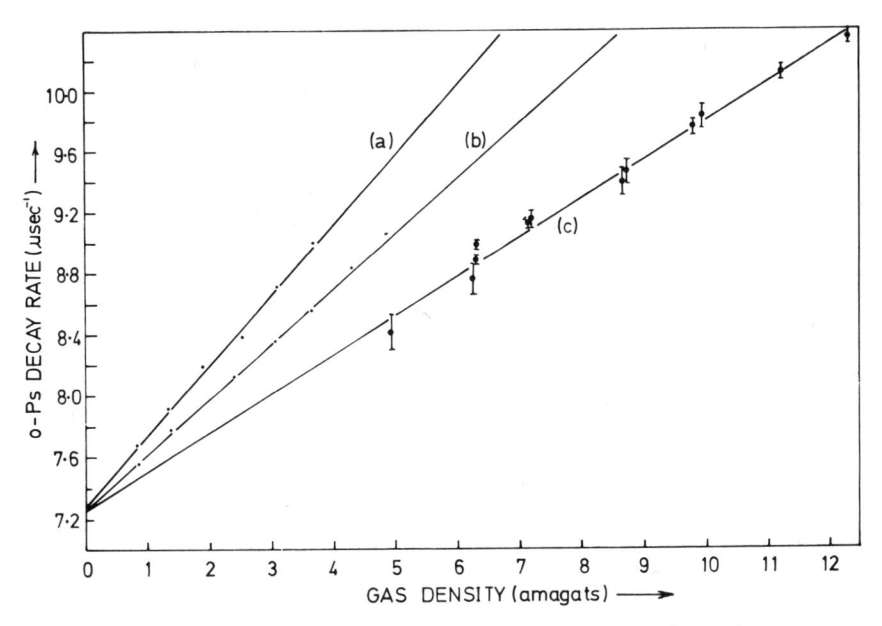

Fig. 11. Observed orthopositronium lifetime in (a) Arcton 12 (b) Freon 12 and (c) argon as a function of gas density.

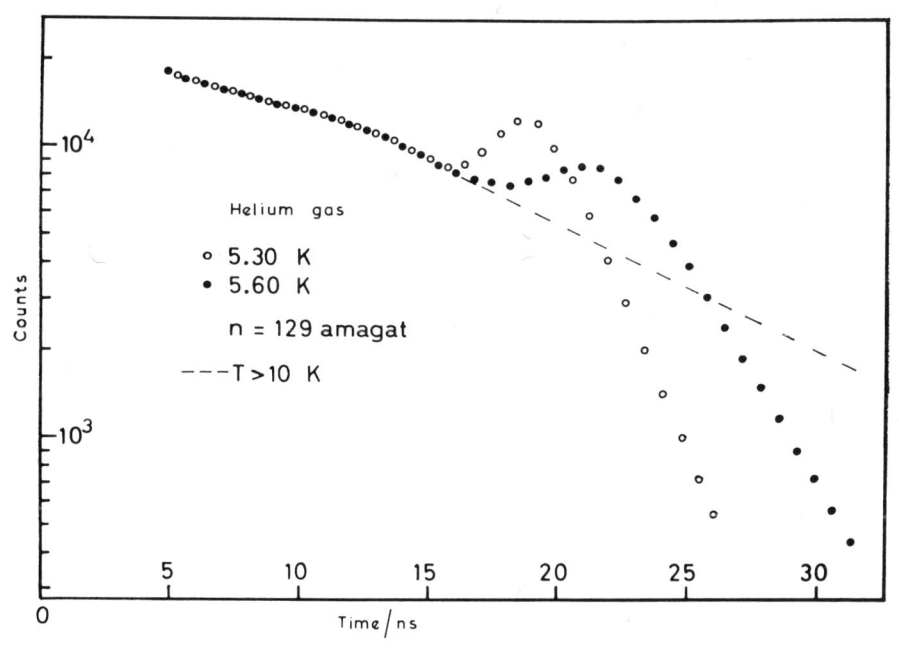

Fig. 12. Observed annihilation spectra of slow positrons in helium gas at low temperatures.

which agrees with an earlier measurement of Theriot, Beers, Hughes and Ziock[6] but is somewhat larger than the theoretical value $7.212 \mu sec^{-1}$. This matter is being further investigated.

Incidentally Arcton 12 and Freon are extremely effective in annihilating free positrons and so, if accurate measurements on the ortho-positronium lifetime are required, it is convenient to work with them as they practically eliminate the free positron component. In the course of the experiments at University College $\bar{Z}_a$ was measured for the first time for Arcton 12 and found to be 480 ± 40 per gas molecule.

Having obtained $\bar{Z}_a$ as a function of E/n it is possible to obtain information about $Z_a(v)$ and $Q_d(v)$. However, this is only possible by trial and error procedures which by no means lead to unambiguous results. The new data on $Q_t(v)$ will provide a valuable check which should reduce the ambiguity. Thus if a theoretical model can be produced which reproduces Q_t, Q_d can be deduced from this with some certainty.

We can hope before long to have a fairly complete analysis for helium to give $Z_a(v)$ accurately from the experimental data. Theory will be required then to reproduce $Z_a(v)$ as well as $Q_t(v)$.

Positron annihilation at low temperatures. Fig. 12 shows annihilation spectra of positrons in helium at the same concentration as in Fig. 8 but at very low temperatures.[7] A new phenomenon is now apparent in that a peak appears just after the shoulder, after which rapid equilibrium decay ensues. At temperatures above about $10°K$ the peak disappears and the spectra cannot be distinguished from those taken at room temperature.

Fig. 13 shows the equilibrium decay rate as a

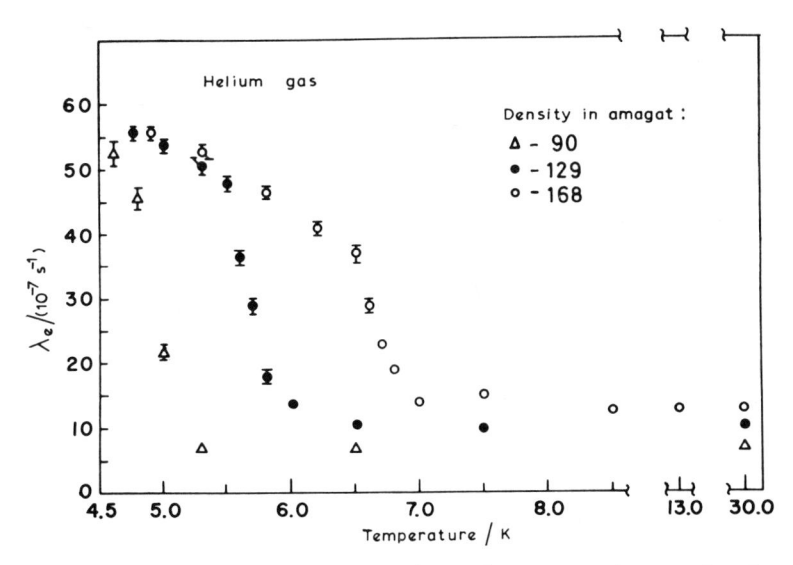

Fig. 13. Observed equilibrium decay rates of slow positrons in helium gas as a function of temperature for three different gas densities.

function of temperature in helium at three different densities.[7] Above 30°K the decay rates agree with those at room temperature whereas at about 4.5°K the decay rate is nearly independent of density and much higher than at room temperature.

At the low temperatures concerned the wavelength of free thermalized positrons is much greater than the atomic separations. It is suggested that the phenomenon is due to cluster formation around a free positron, so greatly increasing the local electron density and hence the annihilation rate.

No effect of this kind is observed in neon[7] as may be seen from Fig. 14. The equilibrium decay rates at densities nearly as high as liquid neon agree well with values obtained by linear extrapolation from the low densities. On the other hand, an effect apparently similar to that in helium is observed in argon[7] at

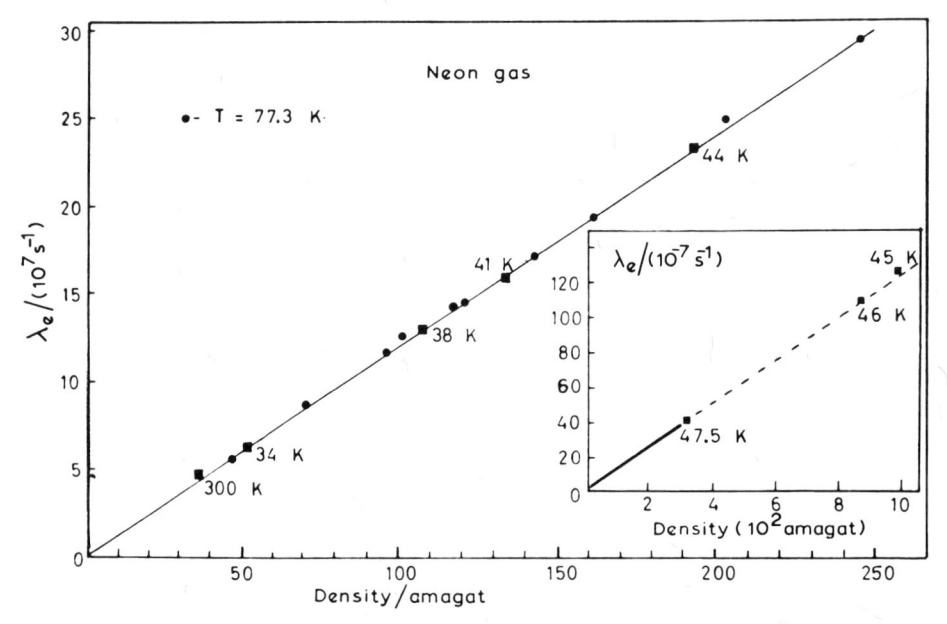

Fig. 14. Observed equilibrium decay rates of slow positrons in neon gas as a function of gas density, including observations at a number of temperatures as indicated.

temperatures below 150°K as may be seen for the spectra shown in Fig. 15. However, in argon under these conditions the positron wavelength is less than the mean atomic distance. It is possible that the peaks shown in Fig. 15 arise from capture of positrons by individual argon atoms and not to cluster formation as in helium.

Observations of this kind offer the possibility of obtaining interesting new information about cluster formation in liquids. It is noteworthy that positronium annihilation in low temperature conditions shows evidence of the opposite phenomenon - bubble formation. There is a net repulsive interaction between orthopositronium and helium.

Concluding remarks. We have attempted here to draw attention to some of the recent developments in the study

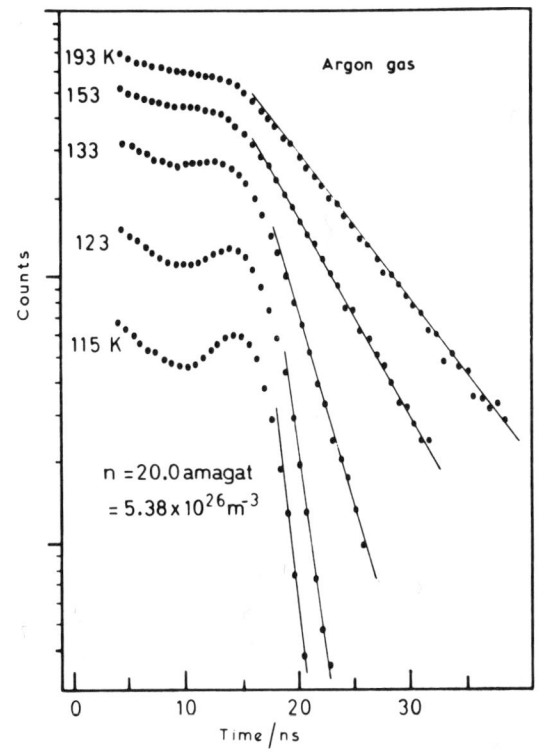

Fig. 15. Observed annihilation spectra of slow positrons in argon gas.

of slow positron collisions in atomic gas. Many aspects of the subject which have been previously reviewed have not been referred to, including almost all of the extensive subject of positronium reactions.

Acknowledgments. I am indebted to the members of the positron research group in my laboratory at University College, London, for the provision of material, permission to present new results and for many valuable discussions.

REFERENCES

1. Humberston J.W. and Wallace, J.B.G., J.Phys.B **5**, 1138-48, 1972.
2. Coleman, P.G., Griffith, T.C. and Heyland, G.R. J.Phys.E **5**, 376, 1972

3. Costello, D.G., Groce, D.E., Herring, D.F. and McGowan, J.W., Phys.Rev.B $\underline{5}$, 1433-6, 1972.
4. Coleman, P.G., Griffith, T.C. and Heyland, G.R., Proc.Roy.Soc.A $\underline{331}$, 561-9, 1973.
5. Canter, K.F., Coleman, P.G., Griffith, T.C. and Heyland, G.R., J.Phys.B $\underline{5}$, L167-9, 1972.
6. Theriot, E.D., Beers, R.H., Hughes, V.W. and Ziock, K.O.H., Phys.Rev.A $\underline{2}$, 707-21, 1970.
7. Canter, K. and Roellig, L.O., Phys.Rev.Lett. $\underline{25}$, 328, 1970; Canter, K. Thesis, Wayne State University, Detroit. 1970.

VARIATIONAL PRINCIPLES, AND ATOMIC SCATTERING

Larry Spruch*

Department of Physics

New York University, New York, New York 10003

I. INTRODUCTION

Since I would like to consider three topics, it will be necessary to simply touch upon each. The three topics center on the development and use of variational principles and variational bounds.

(i) The first problem, perhaps appropriate to a conference that has centered on high energies, is the determination, within the framework of the Schroedinger equation, of the velocity dependence of transfer cross-sections at very high energies. The problem of primary interest is the velocity dependence at high energy of the charge exchange cross-section $\sigma(p + H \to H + p)$, the interest deriving largely from the fact that the <u>second</u> rather than the first Born term seems to be the determining term. The problem has been an outstanding one

---

*Supported in part by the U.S. Army Research Office, Durham, Grant No. DA-ARO-D-31-124-72-G92 and by the Office of Naval Research, Grant No. N00014-67-A-0467-0007.

for more than 40 years,[1] and will unfortunately remain outstanding after my remarks. Using a known variational bound[2,3] for transition amplitudes, and exploiting the known connection[4] between the true problem and the impact parameter approximation, we have, however, found a related example - apparently the first such example - where the second Born term dominates at sufficiently high energies. The work was done in conjunction with - that is a euphemism for largely by - Dr. Robin Shakeshaft.

(ii) It is relatively easy to obtain one variational bound - an estimate that gives an error of known sign and of second order in the error of the input information (for example, the difference $\delta\phi$ between the trial wavefunction ϕ_t and the exact wavefunction ϕ) - on scattering parameters and energies; a beautifully simple variational upper bound on the energy, for example, is provided by the Rayleigh-Ritz theorem. I would like to report on some work done by Professor Yukap Hahn and myself on the possibility of using the adiabatic approximation to obtain the <u>other</u> variational bound; for the energy, for example, this would be a variational lower bound.

(iii) Variational principles for energies and scattering parameters involve only trial wavefunctions of the exact wavefunctions that define the energies and scattering parameters; this is a consequence of the fact that energies and scattering parameters can be expressed as matrix elements of the Hamiltonian H. More generally, however, variational principles involve not only trial functions of the functions that define the quantity of interest but "auxiliary functions".[5,6,7] (These can be thought of as Lagrange undetermined multipliers that account, to first order, for the constraints imposed upon the functions that appear in the definition of the quantity to be estimated.) Gerjuoy, Rau, Rosenberg and I have just obtained what appears to be the first general prescription for converging on the auxiliary functions, and we hope that it will now be possible to greatly expand the domain of realistic applicability of variational principles. We have in mind, for example, the determination of numerical estimates of diagonal and off-diagonal matrix elements of a known linear Hermitian operator with respect to Schroedinger wave functions that are well defined but inexactly known.

2. VELOCITY DEPENDENCE OF EXCHANGE COLLISION CROSS SECTIONS AT HIGH ENERGY[8]

Consider the laboratory frame, with a particle of mass M_a initially at rest and with a particle of mass M_b incident with very high velocity $\vec{v}$. A particle of mass m is initially bound to M_a. It is known[4,9] that in the limit as the ratios m/M_a and m/M_b vanish, the main contribution to the exchange cross section

$$\sigma(M_b + [m + M_a] \to [M_b + m] + M_a)$$

comes from the neighborhood of forward scattering, M_a emerging with some slight recoil velocity and M_b emerging, with m attached, with a velocity close to $\vec{v}$. σ is proportional to v^{-q}, and the problem is to determine q; further, one would like to have some physical insight into the nature of the exchange process.

A considerable effort has gone into this problem. For $\sigma(p + H \to H + p)$, for example, Thomas[10] as far back as 1927 gave a classical argument that led to $q = 11$; the electron, struck by the (almost undeflected) incident proton, bounced off the target proton and emerged with the same velocity as the outgoing proton. Contact between the classical and quantum mechanical treatments has yet to be properly made, but the two-step character of the dominant classical contribution at least suggests that the dominant quantum mechanical contribution might well be that of the second Born term.

Quantum mechanically, ignoring the proton-proton interaction, the first Born term gives[11] $q = 12$. Still in the first Born approximation, the value of q is not affected by the inclusion of the proton-proton interaction.[12] The <u>second</u> Born term gives[13] $q = 11$, in agreement with the classical result. Furthermore,[9] the answer remains $q = 11$ for any finite number $N > 2$ of Born terms, the dominant contribution remaining that of the second Born term. This result does <u>not</u> constitute a proof that $q = 11$; it is well known that the expansion of the full Green's function G about the free Green's function G_o, with V the sum of the pair interactions,

$$G = G_o + G_o V G_o + \ldots$$

does not converge for rearrangement collisions,[14] and there is no reason to assume that the Born expansion for σ converges.

Important additional information about the high energy exchange process, of a rather different character, is provided by the proof[13] of the identity of the N-th term in the Born expansions of the scattering amplitude of the true problem and the transition amplitude A in the impact parameter approximation. In this latter approximation, M_a and M_b are treated classically, M_a remaining at rest and M_b travelling with constant velocity $\vec{v}$, its position being a specific function of the time. m is treated quantum-mechanically, its interaction V(t) with the protons a function of time (as well as of its position). We are now concerned not with a scattering process but with a transition, the transition amplitude being given by

$$A = \lim_{\substack{T_o \to -\infty \\ T \to \infty}} (\Psi_b(T), U(T,T_o)\Psi_a(T_o)) \ . \qquad (2.1)$$

$\Psi_a(T_o)$ is the wavefunction in the remote past, with M_b incident on the bound system $m + M_a$, while $\Psi_b(T)$ is the wavefunction in the far future, with M_a still at rest, with M_b still moving with velocity $\vec{v}$, but with m now attached to M_b. The only unknown in (2.1) is the time-translation operator U which, for finite (and infinite) times, satisfies $U^\dagger U = 1$. We can therefore hope to bound any matrix element containing U by the use of the Schwarz inequality.

(The finiteness of the unknown element U of the transition process is to be contrasted with the singular nature of the unknown element G of the scattering process. It is not surprising that we can much more readily obtain bounds for transition amplitudes than for scattering amplitudes; the nonnormalizable wavefunctions associated with the protons have been eliminated in the impact parameter approximation.)

The use of the Schwarz inequality in a study of p + H → H + p gives a rigorous but - depending upon the details of the approach - a poor or useless result; the singular nature of the Coulomb potential is the source of the difficulty. Consider, however, the case for which all of the potentials are finite. It is then known[9] that the first Born term in the expansion of A leads to $q = 20$, the second to $q = 19$, and the sum of the first N for $N \geq 2$ to $q = 19$; here too, the second Born term dominates over the first and higher Born terms. Now, however, the question of the convergences of the Born expansion can be avoided. We iterate N times the integral equation satisfied by U, about the free time translation operator U_o; suppressing time dependences and time integrations, we write, symbolically,

$$U = U_o + U_o V U_o + \ldots + (U_o V)^{N+1} U \quad ,$$

where V is again the sum of the pair interactions. We can then write

$$A = \sum_{n=1}^{N} A_{nB} + R_N \quad ,$$

where the A_{nB} are the Born terms, which depend upon U_o but not U, and where the remainder, $R_N = R_N(U)$, can be bounded,

$$|R_N| \leq O(v^{-N-1}) \quad .$$

Choosing N to be sufficiently large, we then have $q = 19$; to our knowledge, this is the first scattering problem — for which the direct transition is not forbidden by selection rules — for which it has been shown more or less rigorously that the second Born term dominates.

3. THE ADIABATIC APPROXIMATION AND THE DETERMINATION OF THE OTHER, MORE DIFFICULT BOUND[15]

Since the Hamiltonian H is bounded from below but not from above, we can expect the simple bound on the ground state energy E to be the upper bound, and a beautifully simple upper variational bound on E is indeed provided by the Rayleigh-Ritz theorem. Correspondingly, the possibility of obtaining a simple upper variational bound on the scattering length[16] A has its origins in the fact that A characterizes scattering at zero incident kinetic energy, at the bottom of the continuous spectrum of H.

The above remarks can be restated somewhat more physically, if somewhat more vaguely, as follows - in any approximate calculation, as one uses the potential V more and more "effectively" by choosing successively more flexible trial wavefunctions, the energy estimate is lowered and converges on E, while, for incident kinetic energy zero, the approximate wavefunction is drawn in more and more, so that, by its definition, the scattering length estimate is lowered and converges on A.

A similar remark obtains with regard to the phase shift η. For e^-H scattering for energies below the threshold for excitation, for example, it is known[17] that as one allows the trial function to include more and more hydrogenic states, thereby allowing the atom to more and more effectively adapt itself to the influence of the incident electron in that the effective electron-atom interaction is lowered, the successive phase shift approximations monotonically approach η, that is, with the argument denoting the states included,

$$\eta(\text{static}) \equiv \eta(1s) \leq \eta(1s + 2s) \leq \ldots \leq \eta \quad .$$

The above remarks suggest that an attempt to obtain the other bound should be based on an approximation in which the effective interaction is <u>deeper</u> than the true interaction. Consider, for example, atom-atom scattering for zero incident relative kinetic energy. With $\vec{r}_1$ and $\vec{r}_2$ representing the collections of coordinates of the two atoms, $\vec{R}$ their separation, H_1 and H_2 their Hamiltonians, $T_{\vec{R}}$ the relative kinetic energy operator, and V_{12} the atom-atom interaction, we have

$$H(\vec{r}_1,\vec{r}_2,\vec{R}) = T_{\vec{R}} + H_1(\vec{r}_1) + H_2(\vec{r}_2) + V_{12}(\vec{r}_1,\vec{r}_2,\vec{R}) . \quad (3.1)$$

If we freeze the nuclei at a separation R, and obtain the associated (adiabatic) ground state energy $\varepsilon_{ad}(R)$, we have, by definition,

$$H_1 + H_2 + V_{12} \geq \varepsilon_{ad}(R) ; \quad (3.2)$$

it follows at once that

$$H \geq T_{\vec{R}} + \varepsilon_{ad}(R) \equiv H_{ad}(R) \quad . \quad (3.3)$$

Physically, the inequality (3.3) is perhaps to be expected, since the electrons now have an infinite amount of time to (over) adapt to one another and to the nuclei. To simplify the discussion I will ignore the problems raised by the fact that H and H_{ad}, as defined above, operate in the different spaces $(\vec{r}_1,\vec{r}_2,\vec{R})$ and $\vec{R}$, respectively. The argument that follows is then not quite honest, but it contains "the physics" of the situation, and the argument can be given correctly.

VARIATIONAL PRINCIPLES

To go beyond reasonability arguments, we proceed as follows. The determination of the usual upper variational bound on A starts from the identity

$$A = A_t + (\Phi_t, [H-E_o]\Phi_t) - (\delta\Phi, [H-E_o]\delta\Phi) \qquad (3.4)$$

as a starting point. E_o is the sum of the ground state energies of the isolated systems and $\psi_o(\vec{r}_1, \vec{r}_2)$ is the product of the ground state wave functions

$$H\Phi = E_o\Phi, \quad \Phi \longrightarrow \text{const}(R-A)\psi_o, \qquad (3.5)$$

H_t is a trial Hamiltonian,

$$H_t\Phi_t = E_o\Phi_t, \quad \Phi_t \longrightarrow \text{const}(R-A_t)\psi_o, \qquad (3.6)$$

and $\delta\Phi = \Phi_t - \Phi$. The identity (3.4) follows immediately from (3.5) and (3.6). Since we are interested in the other bound, we interchange the roles of H and H_t, and choose $H_t = H_{ad}$ and $\Phi_t = \Phi_{ad}\psi_o$, where then

$$H_{ad}\Phi_{ad} = E_o\Phi_{ad}, \quad \Phi_{ad} \longrightarrow \text{const}(R-A_{ad})\psi_o. \qquad (3.7)$$

We then have

$$A_{ad} = A + (\Phi, [H_{ad}-E_o]\Phi) - (\delta\Phi, [H_{ad}-E_o]\delta\Phi) \equiv A + B - C. \qquad (3.8)$$

Replacing $E_o\Phi$ by $H\Phi$ and using $H_{ad} - H \leq 0$, it follows that $B \leq 0$. If H_{ad} cannot support any bound states, $H_{ad} - E_o$ is non-negative with respect to functions which are quadratically integrable or[16] the limits of quadratically integrable functions, such as $\delta\Phi$ (but _not_ Φ).

We then deduce that $C \geq 0$ and therefore that

$A_{ad} \leq A$, if H_{ad} cannot support bound states.

The inequality remains valid if H_{ad} can support bound states, provided H can support the same number. [An analogous restriction exists[18] for the validity of the other inequality, $A \leq A(\text{static})$.]

The adiabatic approach therefore often provides the difficult (lower) bound on A. The method can readily be extended to provide the difficult (upper) bound on the elastic scattering phase shift η. It has been known for some time that, as follows from the above arguments, the adiabatic approximation provides the difficult (lower) bound on the ground state energy. What Hahn and I - in fact, mostly Hahn - are trying to do is use the above approach to obtain a <u>variational</u> expression for the difficult bounds on the ground state energy, A, and η.

4. AN IMPROVEMENT IN THE <u>USE</u> OF VARIATIONAL PRINCIPLES[19]

To simplify the discussion, we consider the special but very important problem of the use of a variational principle to estimate the diagonal matrix element $\langle W \rangle \equiv (\phi, W\phi)$ for W a linear Hermitian operator and φ a real normalized ground state wavefunction. The basic results were originally derived somewhat differently,[5,6,7] but from our point of view the best way to look at the problem is to recognize that the equation $(H-E)\phi = 0$ represents a constraint at each point $\vec{r}$, requiring a Lagrange multiplier at each point $\vec{r}$ and a summation, or rather an integration, over all points $\vec{r}$. For notational convenience, we introduce the Lagrange multiplier $L_t(\vec{r})$, and, with ϕ_t normalized, we write

$$<W>_{var} = (\phi_t, W\phi_t) + 2(L_t, [H-E_t]\phi_t) \quad , \quad (4.1)$$

where $E_t \equiv (\phi_t, H\phi_t)$. However we motivate (4.1), it will represent a variational principle if and only if we can define our Lagrange multiplier or auxiliary function $L_t(\vec{r})$ so that the first order variation of $<W>_{var}$ vanishes. The requirement that $\delta<W>_{var} = 0$ leads to

$$(H-E)L = (\phi, W\phi)\phi - W\phi \equiv f(\phi) \quad (4.2)$$

as the defining equation for L.[5,6,7] A first order estimate L_t of L must be obtained if (4.1) is to be useful. We note that the requirements that

$$(\phi, L) = 0 \quad (4.3)$$

and that L be quadratically integrable make L unique.

It is well known[20] that one cannot simply replace (4.2) by

$$(H-E_t)L_t = f(\phi_t) \quad ; \quad (4.4)$$

the use of (4.4) in (4.1) gives the crazy result $<W>_{var} = (\phi_t, W\phi_t)$. An understanding of the origin of the loss of the variational principle, which had not previously been given, is essential if one is to improve upon the normal practice. The normal practice is to replace (4.2) and (4.3) by

$$(\hat{H}-\hat{E})\hat{L} = f(\hat{\phi}) \quad (4.5)$$

$$(\hat{\phi}, \hat{L}) = 0 , \qquad (4.6)$$

where $\hat{H}$ is a solvable approximation to H,

$$(\hat{H}-\hat{E})\hat{\phi} = 0 . \qquad (4.7)$$

The restriction to solvable approximations $\hat{H}$ is extremely undesirable; I will now discuss briefly how to eliminate this restriction, while insisting that $H-E$ be replaced by a non-negative operator.

In (4.4), there is no solution of the homogeneous equation that satisfies the appropriate boundary conditions and L_t is uniquely defined; we cannot impose an orthogonality condition upon ϕ_t analogous to the orthogonality condition (4.3) imposed upon ϕ. As ϕ_t approaches ϕ, E_t approaches E; the inversion of $H-E_t$ necessary to the determination of L_t therefore generates a component of L_t proportional to $(E-E_t)^{-1}$, which approaches infinity. This problem is avoided in the determination of L of (4.2) by the condition (4.3) and in the determination of $\hat{L}$ of (4.5) by the condition (4.6). It is possible, in a general approach allowing an arbitrary ϕ_t, to eliminate the near singularity associated with the inversion of the approximation to $(H-E)$ by adapting techniques that arose in the determination of a variational bound on the scattering length[16] and in work on the effective potential.[21] Thus, for ϕ_t fixed, we replace (4.2) by

$$O_t L_t \equiv [H_{mod,t} - E_t(1-P_t)]L_t = f(\phi_t) , \qquad (4.8)$$

where again $E_t = (\phi_t, H\phi_t)$ and where the trial modified Hamiltonian is defined by

VARIATIONAL PRINCIPLES   225

$$H_{mod,t} \equiv H - \frac{H P_t H}{E_t} \quad , \tag{4.9}$$

with P_t a projection operator, $P_t = \phi_t \phi_t^\dagger$. To simplify the discussion, we assume that the system under consideration has only one bound state, and we assume further that ϕ_t is good enough to give binding, that is, that $E_t < 0$. We now make the following observations.

i) ϕ_t is a solution of the homogeneous equation associated with (4.8), that is $O_t \phi_t = 0$. Since $(\phi_t, f) = 0$, a solution L_t of (4.8) exists.

ii) Imposing the condition

$$(\phi_t, L_t) = 0 \tag{4.10}$$

eliminates any singularity or near singularity difficulties.

iii) As ϕ_t approaches ϕ, O_t approaches H-E. There is therefore no limit, in principle, to the accuracy that can be achieved.

iv) $H_{mod,t}$ is known to be a non-negative operator,[16,21] the negative energy contribution associated with the one bound state having been "extracted" even without a knowledge of the bound state wavefunction. Since $E_t(1-P_t)$ is also clearly non-negative, it follows that $O_t \geq 0$. This property enormously simplifies the task of obtaining an approximation L_{tt} to L_t, which in turn is an approximation to L. [For a fixed ϕ_t, the best we can do is obtain L_t, defined by (4.8) and $(\phi_t, L_t) = 0$.] Thus, given $O_t L_t = f(\phi_t)$, with $f(\phi_t)$ known and $O_t \geq 0$, it is well known[22] that one can construct a functional $M(L_{tt})$ which achieves its extremum value for $L_{tt} = L_t$; this gives us a powerful prescription for choosing any variational parameters contained in L_{tt}.

The above approach can readily be generalized to the case of an arbitrary number of bound states, to off-diagonal matrix elements involving complex wavefunctions, and to the extremum evaluation of a Green's function defined by $(H-E)G = (1-P)$, where E is an eigenvalue of H and P is the associated projection operator.

REFERENCES

1. See, for example, C.S. Shastry, A.K. Rajagopal and J. Callaway, Phys.Rev. A6, 268 (1972), B.H. Bransden, Rep. Prog. Phys. 35, 949 (1972), and R.A. Mapleton, Theory of Charge Exchange (Wiley, New York, 1972).
2. L. Spruch, Lectures in Theoretical Physics, Vol. XI c, p. 77, S. Geltman, K.T. Mahanthappa and W.E. Brittin, eds. (Gordon and Breach, New York, 1969).
3. L. Aspinall and I.C. Percival, J.Phys.B 1, 589 (1968) had earlier obtained a bound.
4. L. Wilets and S.J. Wallace, Phys.Rev. 169, 84 (1968). See also N.F. Mott, Proc.Camb.Phil.Soc. 27, 553 (1931)
5. A. Dalgarno and A.L. Stewart, Proc.Roy.Soc. A257, 534 (1960).
6. C. Schwartz, Ann.Phys.(N.Y.), 6, 156 and 170 (1959).
7. L.M. Delves, Nucl.Phys. 41, 497 (1963).
8. The material in this section is based entirely upon work with Shakeshaft: R. Shakeshaft and L. Spruch, submitted for publication.
9. K. Dettmann and G. Leibfried, Z.Physik 218, 1 (1969); K. Dettmann, Springer Tracts in Modern Physics, 58, 119 (1971).
10. L.H. Thomas, Proc.Roy.Soc.A 114, 461 (1927).

11. H.C. Brinkmann and H.A. Kramers, Proc.Acad.Sci. (Amsterdam) 33, 973 (1930).
12. J.D. Jackson and H. Schiff, Phys.Rev. 89, 359 (1953).
13. R.M. Drisko, Thesis, Carnegie Institute of Technology (1955).
14. R. Aaron, R.D. Amado, and B.W. Lee, Phys.Rev. 121, 319 (1961).
15. The material in this section is based entirely upon work with Y. Hahn which is to be submitted for publication.
16. L. Spruch and L. Rosenberg, Phys.Rev. 116, 1034 (1959); L. Rosenberg, L. Spruch, and T.F. O'Malley, Phys.Rev. 118, 184 (1960).
17. Y. Hahn, T.F. O'Malley, and L. Spruch, Phys.Rev. 128, 932 (1962).
18. L. Spruch and L. Rosenberg, Nucl.Phys. 17, 30 (1960).
19. The material in this section is based entirely upon a paper by Gerjuoy, Rau, Rosenberg and the author which will shortly be submitted for publication.
20. J.B. Krieger and V. Sahni, Phys.Rev. A6, 919 (1972).
21. W. Monaghan and L. Rosenberg, Phys.Rev. A6, 1076 (1972).
22. See, for example, G.S. Michlin, *Variational Methods in Mathematical Physics* (McMillan, New York, 1964).

PANEL DISCUSSION

SPIN AND POLARIZATION EFFECTS IN ATOMIC PROCESSES

>
> Panel Chairman: H. Kleinpoppen
> (University of Stirling)
> Panel Members: B. Bederson (New York University), P. Lambropoulos (J.I.L.A.), J. Macek (University of Nebraska), H.S. W. Massey (University College, London), E. Reichert (Universität Mainz)

KLEINPOPPEN (Introductory paper)

In view of all the present activity on spin and polarization effects in atomic processes, perhaps the best approach to my introduction is first to summarize some of the most important recent results and to restrict myself to some typical characteristics.

To begin with, let me divide the field to be discussed into atomic collision and atomic photoionization processes. As will be seen in the course of discussion, one of the most important recent applications of spin effects consists of using electronic and atomic spin polarization in order to separate different types of electromagnetic interactions from each other. It seems to me that the present new scattering experiments with polarized electrons and polarized atoms may well develop into a classical example of how different types of interactions involved in a given process can be studied separately.

As you remember, the first theoretical studies on electron spin polarization by scattering from unpolarized targets were carried out for high energy scattering by Mott[1] around the 30's, and at the beginning of the 40's by Massey and Mohr[2] for low energy scattering. The term high and low energy refers either to electron energies high enough for electrons to pass the atomic shells and where the electrons experiences a pure Coulomb field, or alternatively the term "low energy" refers to energies too low so that the electrons are scattered by a strongly screened Coulomb field.

Meanwhile, since the early investigations of Mott and Massey and Mohr, the theory and also experimental data taken on spin polarization of low and high energy scattering from unpolarized heavy target atoms like Hg reached a rather advanced status. The polarization at high energies of about 100 keV mainly results from spin orbit interactions of the electrons in the pure Coulomb field whereas at low energy exchange effects may also play an important role. Dr. Reichert, our panel member, who is one of the experimental experts and pioneers in this field, may be referring to the latest developments. In this connection, I only would like to pick up an extremely interesting example for electron resonance scattering which links very nicely with analogous experiments in particle and nuclear physics, viz. that the neutron or proton scattering in He shows anomalies for energies in the MeV range which has been attributed to $P_{1/2}$ and $P_{3/2}$ Breit-Wigner resonances associated with the formation of unstable compound states of He^5 or Li^5. This was first predicted by Schwinger[3] in 1946. Due to the large energy splitting between the $P_{1/2}$ and $P_{3/2}$ resonances, the neutrons and protons scattered in these resonances are subjected to a strong spin orbit interaction which manifests itself in the polarization of the elastically scattered neutrons or protons as demonstrated in experiments like that of Heusinkveld and Freier[4] and others[5].

Dr. Reichert and his colleagues[6] studied the polarization of electrons elastically scattered in Ne atoms in an energy range where $P_{1/2, 3/2}$ resonances occurred according to an interpretation of Simpson and Fano[7]. In their apparatus they used an electron beam of 1.10^{-8} A to 3.10^{-8} A with an energy half-width of $\Delta E = 30$ meV to 45 meV from a 180° spherical electron monochromator. The electrons scattered by an atomic beam of neon were analyzed for their intensity and polarization under different angles. Fig. 1 shows a typical result which agrees well with theoretical predictions based upon a theory of Fano and Simpson[7] and Thomson[8].

TABLE I

Types of Interaction	Z small*	Z large*
1) Coulomb or direct, f	✓	✓
2) exchange, g	✓	✓
3) spin-orbit, h	-	✓
4) spin-spin,	-	-
5) spin-nuclear spin,	-	-

*These columns indicate when the different types of interactions are normally taken into account in theoretical calculations (Z atomic number of the target atom). Interference terms of the amplitudes f, g, h are not listed in the table.

Next I would like to discuss how different types of electromagnetic interactions can be separated from each

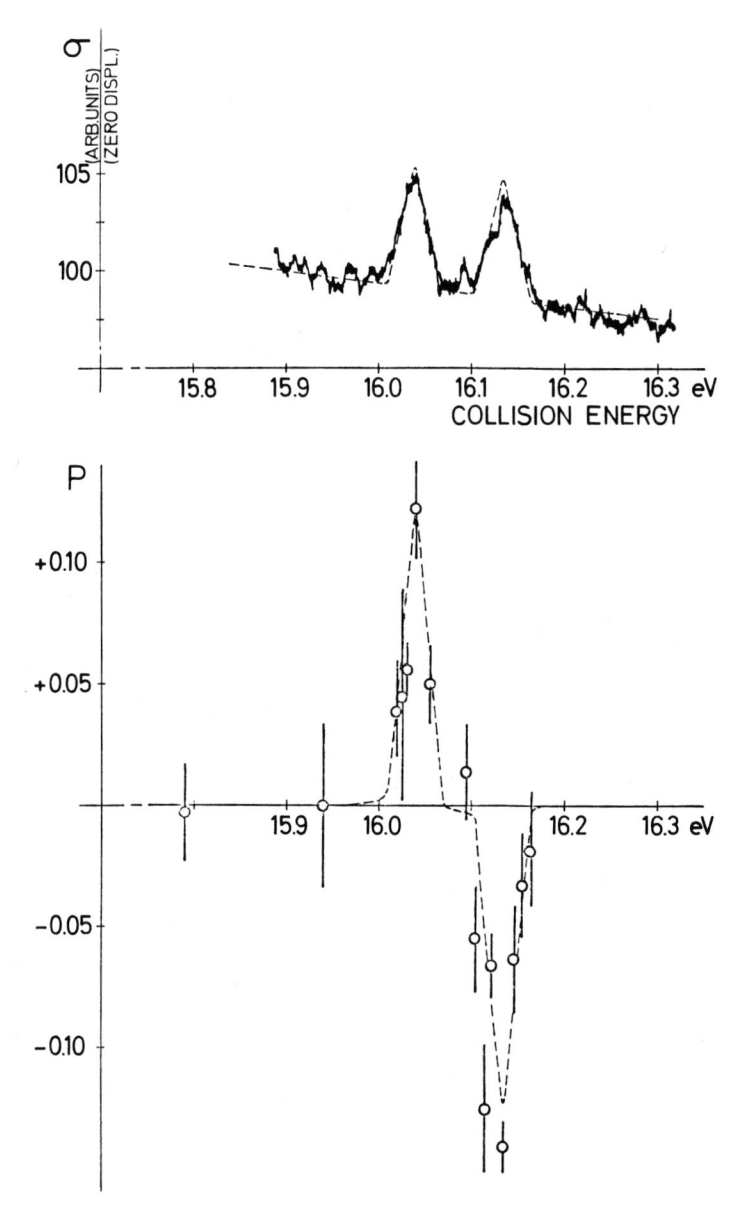

Fig. 1 Electron scattering from Neon[6] in the region of the $P_{3/2}$ (at 16.04 eV) and $P_{1/2}$ resonance, scattering angle $90°$. Upper part: differential cross section, the scattered current at 15.9 eV is set to 100 units.
———— recorder trace, - - - - - theoretical values from theory[7], folded with a triangular profile of half width 30 meV. Lower part: spin polarization of electrons from Neon.

SPIN AND POLARIZATION EFFECTS

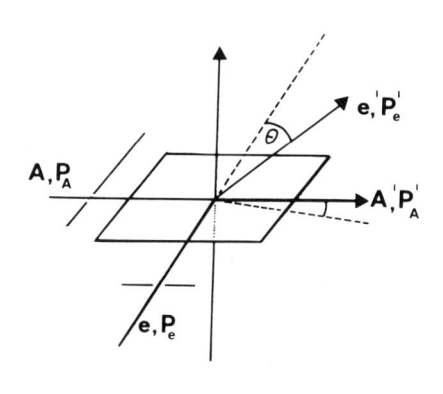

1) $e(\uparrow) + A(\downarrow) \longrightarrow A(\downarrow) + e(\uparrow)$, $|f|^2$
2) $\longrightarrow A(\uparrow) + e(\downarrow)$, $|g|^2$
3) $e(\uparrow) + A(\uparrow) \longrightarrow A(\uparrow) + e(\uparrow)$, $|f-g|^2$

$$\sigma(\theta) = \tfrac{1}{2}|f|^2 + \tfrac{1}{2}|g|^2 + \tfrac{1}{2}|f-g|^2$$
$$= \tfrac{1}{4}|^1F|^2 + \tfrac{3}{4}|^3F|^2, \; ^1F = f+g, \; ^3F = f-g$$
$$= \sigma_d(\theta) + \sigma_{ex}(\theta) + \sigma_{int}(\theta)$$

(a) partially polarized electrons scattered on unpolarized atoms:

$$\frac{|f|^2}{\sigma} = 1 - \frac{P_A'}{P_e}, \qquad \frac{|g|^2}{\sigma} = 1 - \frac{P_e'}{P_e}$$

(b) unpolarized electrons scattered on partially polarized atoms:

$$\frac{|f|^2}{\sigma} = 1 - \frac{P_e'}{P_A}, \qquad \frac{|g|^2}{\sigma} = 1 - \frac{P_A'}{P_A}$$

(c) partially polarized electrons scattered on partially polarized atoms:

$$|f-g|^2 = \sigma + \frac{1}{P_e P_A}(S - \sigma) = |^3F|^2$$
$$\cos\delta = \frac{\sigma - S}{P_e P_A |f||g|}$$

Fig. 2 The upper part illustrates a cross-beam experiment with partially polarized electrons (initial and final polarizations P_e and P_e') and partially polarized atoms (initial and final polarizations P_A and P_A'). The three basic scattering processes (1 - 3) with completely polarized electrons and atoms are associated with the scattering amplitudes f, g, and f - g. The analysis of scattering amplitudes with partially polarized electrons and atoms is illustrated in sections a to c (σ differential cross section, S scattering intensity of partially polarized electrons scattered on partially polarized atoms, δ phase difference between f and g, ^1F ^3F singlet and triplet scattering amplitude).

other by using polarized electrons and polarized atoms in a collision experiment. Table I lists possible types of electromagnetic interactions occuring during an electron atom collision process. Only the first 2 or 3 types of interactions are normally taken into consideration from the remaining ones listed in the table.

Fig. 2 illustrates how an analysis of spin polarization effects in elastic polarized electron atom scattering experiments for one-electron atoms can be used in order to determine the moduli of the amplitudes f, g and f - g for light alkali atoms (neglecting spin orbit interactions!). The upper part of Fig. 2 associates the amplitudes involved with reactions of both completely polarized electrons and atoms. The lower part of the Fig. 2 illustrates the analysis of a crossed beam collision experiment where at least one of the collision partners is partially polarized. As follows from Fig.2 measurements of either the polarization of scattered atoms or electrons supplement each other if one of the collision partners is unpolarized. The measurement of the intensity S for the scattering of partially polarized electrons on partially polarized atoms provides both a means to determine the modulus of the interference amplitude f-g (or the triplet amplitude ^3F!) and also the phase difference δ between f and g. A theoretical analysis of scattering amplitudes including spin orbit interactions, is being investigated by Burke and Mitchell[9], and Blum and Kleinpoppen[10].

To my knowledge there are six experimental groups at present investigating scattering processes with polarized electrons and atoms: at the Universities of Colorado (Joint Institute for Laboratory Astrophysics) Boulder, Edinburgh, Mainz, New York University, Yale and also at Stirling. Fig. 3 shows the scheme of the aparatus used for the successful measurement of differential exchange cross sections (K. Rubin et al[11]). An

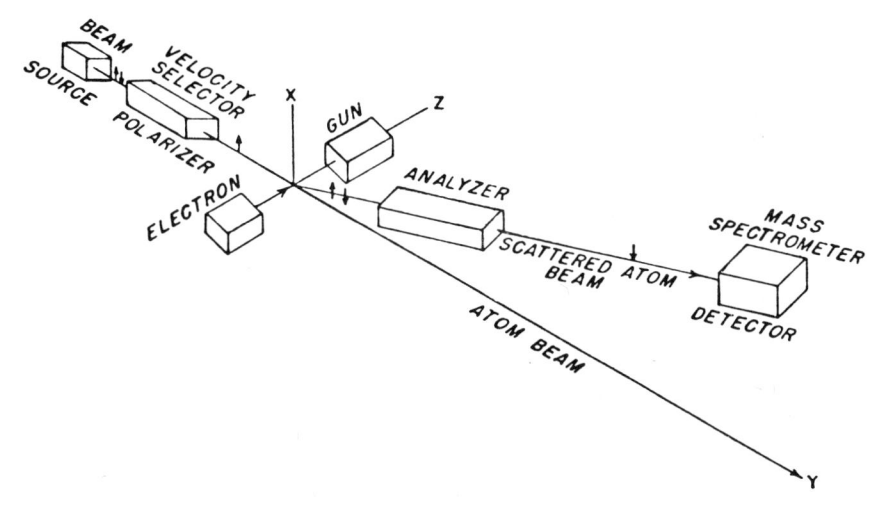

Fig. 3 Schematic diagram of the electron atom recoil experiment with initially polarized alkali atoms[11-13] (NYU group). ↑ indicates direction of spin-polarization.

alkali beam is velocity - and spin-state selected by Stern-Gerlach magnet (two-wire type magnet), before being cross-fired by an unpolarized beam of electrons. After scattering the atom is spin analyzed by an E-H gradient balance magnet[12]. The entire analyzer-detector assembly rotates about the scattering centre in the X-Y plane. The electron gun is rotatable in the X-Y plane about the Y axis. The detector is also translatable in ± Z direction. Using simple kinematic relations the electron polar scattering angle can be obtained from the atomic recoil scattering angle. Normally the polarizer and analyzer are set to transmit opposite spin states so that only spin exchanged (or spin flipped) atoms can reach the detector when the analyzer is operative.

Differential elastic exchange cross sections for potassium have been obtained by Collins et al[13] using this technique. These are obtained from measurements of $R(\Theta) \equiv \sigma_{ex}(\Theta)/\sigma(\Theta)$, combined with direct measurements of $\sigma(\Theta)$ normalized to the total cross section at 1 eV. $R(\Theta)$ is simply the ratio of atom beam current at a given detector position with and without the spin analyzer operative, corrected for transmission and residual depolarization. Partial depolarization caused by hyperfine coupling is avoided in this work by the use of a high (>1000 Gauss) magnetic field in the interaction region. Fig. 4 shows these results, compared to Karule's[14] calculation. The bump in the 1.2 eV experi-

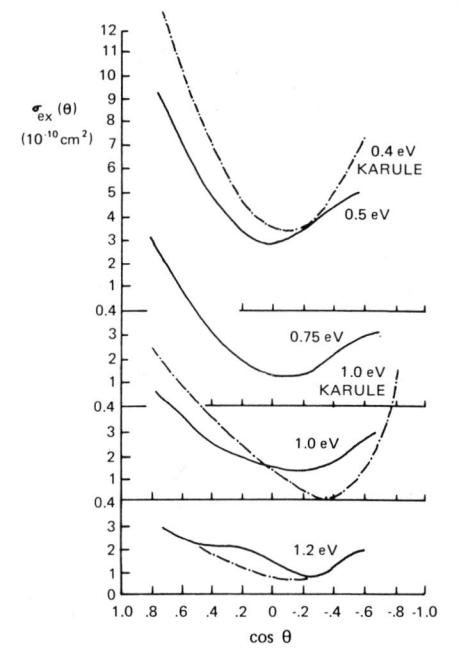

Fig. 4 Differential spin-exchange cross sections for electron potassium scattering at electron energies of 0.5, 0.75, 1.0 and 1.2 eV. Dot-dashed curves from Karule's[14] theory, and solid curves from experiments[13].

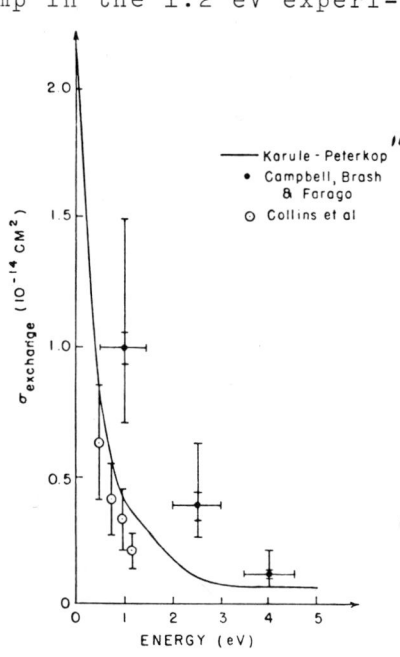

Fig. 5 Total spin exchange cross section for elastic electron potassium scattering versus electron energy.

mental curve in the vicinity of 90° is attributable to spin-flip caused by inelastic scattering from the high-energy tail of the electron energy distribution.

Campbell, Brash and Farago[15] at Edinburgh University used an electron trap in which the polarization of an alkali beam is transferred by exchange processes to initially unpolarized trapped electrons; the exchange collisions gradually polarize the trapped electrons to a maximum polarization equivalent to that of the initial atomic polarization. After reaching the maximum polarization the trapped electrons are subsequently extracted to form a polarized beam which is Mott analyzed. From the degree of polarization as measured by the Mott analyzer the total exchange cross section can be measured. Fig. 5 includes typical data from the Edinburgh group compared with experimental data of the NYU group[13] (obtained by integrating results like those of Fig. 4 over all angles) and with the theory of Karule and Peterkop[16].

Fig. 6 shows the scheme of the J.I.L.A.-Stirling type experiment[17] for the measurement of the elastic direct differential cross section of alkali atoms. A beam of potassium atoms is polarized by the field of a hexapole magnet, and the polarization is aligned parallel to the atomic beam direction by a weak magnetic field (a few milligauss). An unpolarized beam of electrons is cross-fired and the polarization of the scattered beam is measured by a Mott detector. From the measured polarization were derived the relative elastic differential direct cross sections for a given energy. Fig. 7 presents the first pioneering data on the measurement of $\frac{|f|^2}{\sigma}$ for potassium atoms.

Before I am going to mention briefly some of the exciting data on polarization effects in atomic photo-

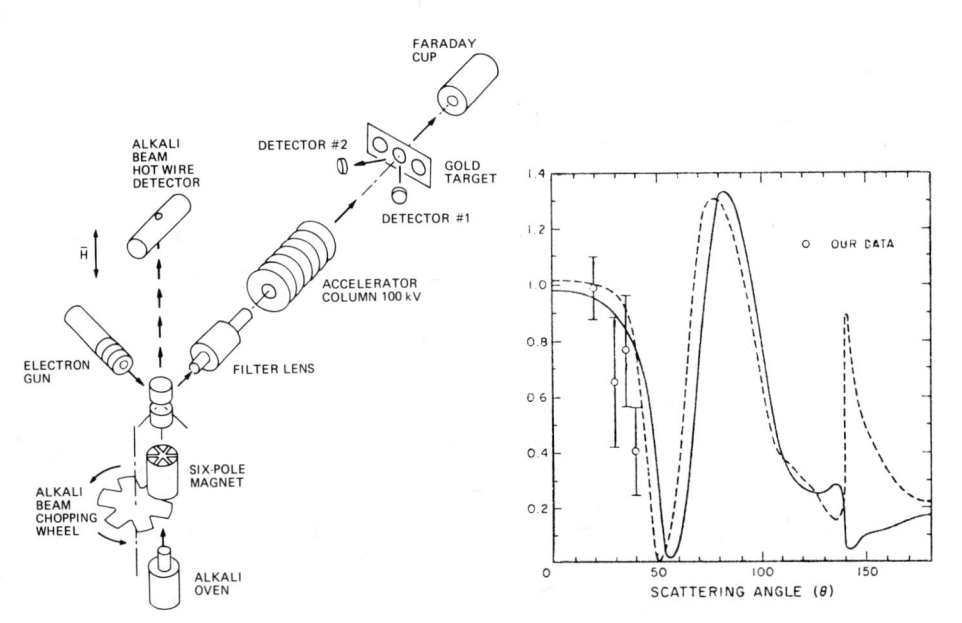

Fig. 6 Schematic diagram of the J.I.L.A.-Stirling type apparatus for the measurement of the "direct" cross section of electron potassium scattering[17].

Fig. 7 Plot of $|f|^2/\sigma$ measurements[17] and theory[16] for electron-potassium scattering: electron energy 3.3 eV in experiment; continuous and dashed curves, theoretical values for 3 and 4 eV, respectively.

ionization processes, I should like to put some emphasis on a result which clearly was expected on theoretical grounds a long time ago. It could, however, not be shown up until very recently because of lack of relevant theoretical and experimental data. Suppose you excite alkali resonance lines by linearly polarized light (see Fig. 8). The resonant florescent light, usually observed perpendicular to the direction of the incoming light, is then also polarized. The polarization can be calculated based upon Breit's theory on resonance florescence, by taking into account the selection rule $\Delta m_\ell = 0$ associated

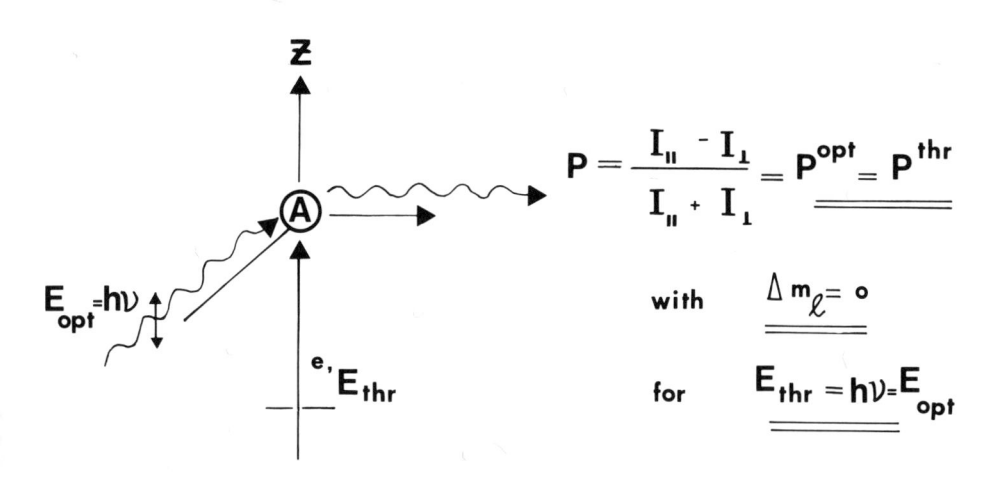

Fig. 8 Equivalance of threshold impact excitation by unidirectional electrons, and resonance excitation by linearly polarized light (z axis parallel to the electron beam direction and the light vector).

with the excitation by linearly polarized light. Now, one of the problems left in collisional excitation of atoms by electrons is whether we can achieve $\Delta m_\ell = 0$ excitation. The answer is yes. You have to use unidirectional electrons of energies corresponding to the threshold energy of the excitation process. The explanation is as follows: The incoming electrons have no component of orbital angular momentum parallel to the z axis. After the excitation at threshold, the electrons have lost all their energy, and accordingly they cannot carry away any orbital angular momentum for compensation of angular momentum transfer, or in other words, the selection rule $\Delta m_\ell = 0$ should hold at the excitation threshold. This rule is verified from the theoretical and experimental data (see Fig. 9).

The table in Fig. 9 displays the experimental and

$Na^{23}, I=3/2$ $Li^6, I=1$ $Li^7, I=3/2$

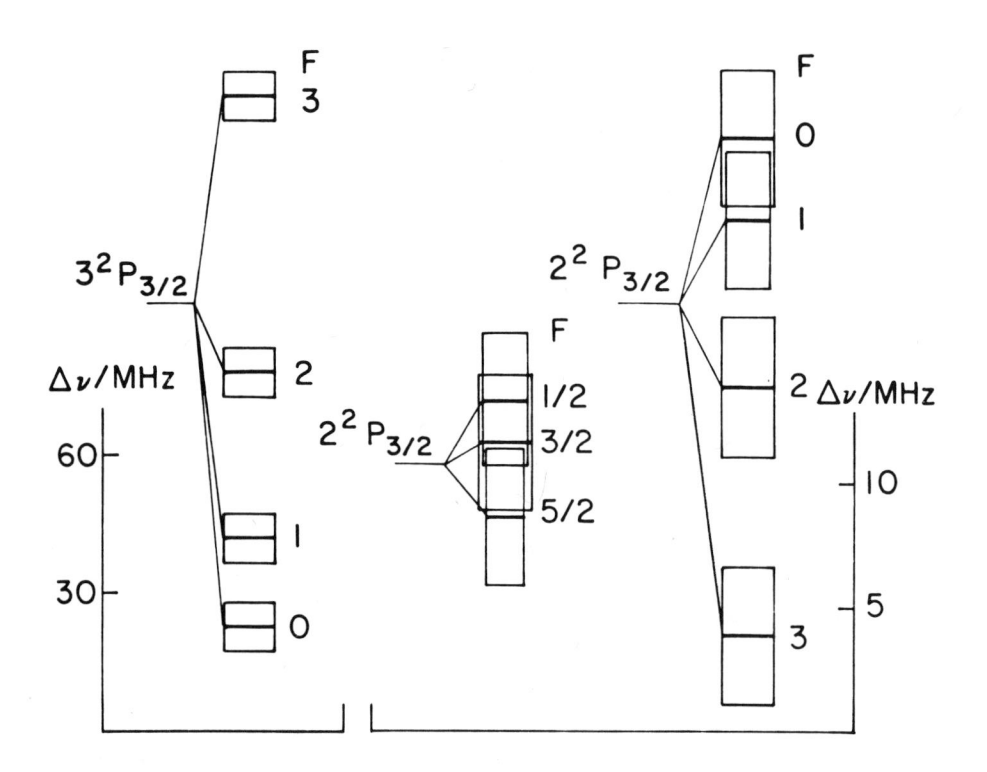

$3^2P_{3/2}$ — F: 3, 2, 1, 0

$2^2P_{3/2}$ — F: 1/2, 3/2, 5/2

$2^2P_{3/2}$ — F: 0, 1, 2, 3

$\Delta\nu$/MHz (left axis: 30, 60)

$2\,\Delta\nu$/MHz (right axis: 5, 10)

$P_{cal}^{thr} = 14.1\%$	37.5%	21.6%
$P_{exp}^{thr} = (14.8 \pm 1.8)\%$	$(39.7 \pm 3.8)\%$	$(20.6 \pm 3.0)\%$
$P_{cal}^{opt} = (14.2 \pm 0.2)\%$	$(37.7 \pm 1.0)\%$	$(21.7 \pm 1.5)\%$
$P_{exp}^{opt} = (14.0 \pm 0.8)\%$	$(37.5 \pm 1.7)\%$	$(23.1 \pm 1.3)\%$

Fig. 9 The upper part of the figure demonstrates the influence of the hyperfine structure separations, and the level width of the first P states on the polarization of the first resonance lines ^{23}Na, ^{6}Li and ^{7}Li. The table in the lower part displays the polarization data as described in the text.

calculated polarization data for the optical and electron threshold excitation of the reasonance line of the Li^6, Li^7 and Na^{23} (P_{cal}^{opt}, P_{exp}^{opt} calculation[18] and experimental[18] polarization data by optical excitation, P_{cal}^{thr}, P_{exp}^{thr} calculated[19] and experimental[20] polarization data by threshold excitation). Note the excellent agreement between the optical and the threshold excitation data on the one hand and also between the theoretical and experimental data. This clearly shows that, at least for light alkali atoms, the two excitation processes under discussion are governed by the same selection rule $\Delta m_\ell = 0$. It also proved for the first time the validity of the theory of polarization of impact radiation[20].

Now I should like to report on two special examples

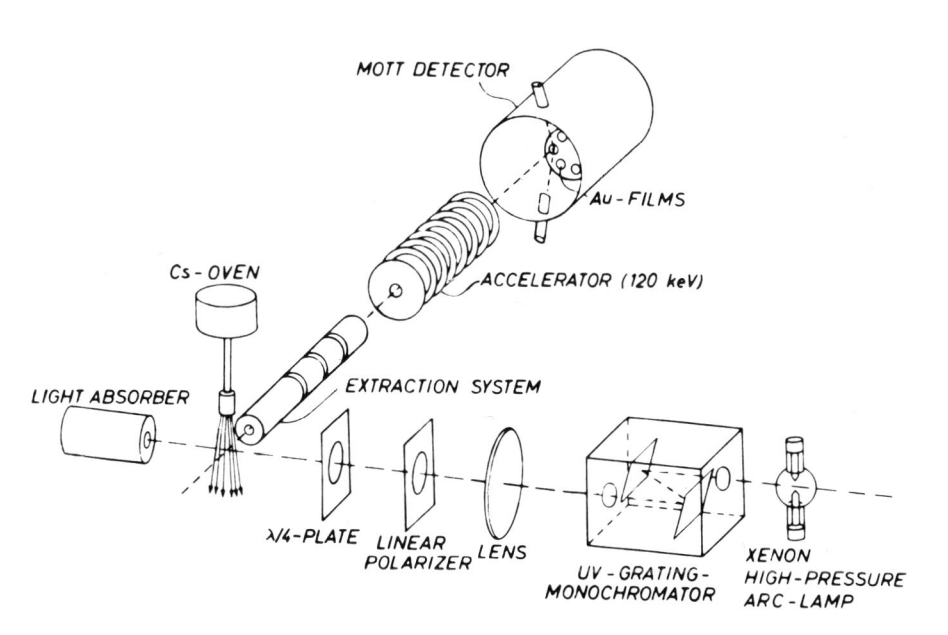

Fig. 10 Scheme of an experiment[23] to test the Fano effect[22]: the polarization of photoelectrons released by circularly polarized light from unpolarized cesium is measured.

for polarization effects in atomic photoionization processes, firstly, on the Fano[21] effect, and secondly on multi-photon ionization processes. The Fano effect can be characterized as an effect physicists almost forgot or overlooked to discover. Let me describe the effect by showing the scheme of the apparatus Fig. 10, which is most suitable for its direct test (Heinzmann et al[23]), although an indirect or modified test version of Fano's original proposal was first given by Baum, Lubbel and Raith[24]. The Fano effect concerns electron spin polarization due to photoionization of atoms by circularly polarized light. A circularly polarized beam of monochromatic light crosses a beam of cesium atoms, the photoelectrons extracted perpendicular to the previous beam direction can be highly polarized as detected by a Mott analyzer, and as shown in Fig. 11 as a function of the wave length of the incoming photons.

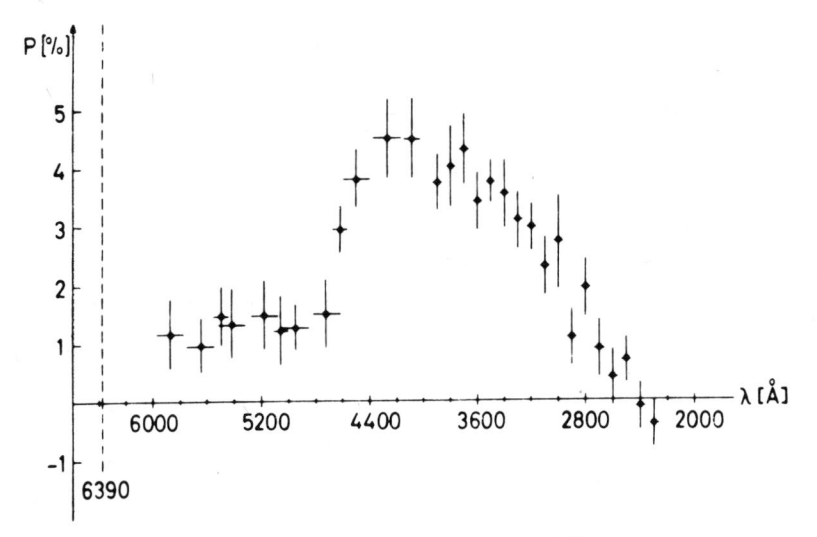

Fig. 11 Detection of the Fano effect[22]: polarization of photoelectrons from unpolarized cesium atoms as a function of the wave length of the incident circularly polarized light[23]. Solid curve: Fano's[22] theoretical prediction.

What is the physical explanation of the Fano effect? As we now know the Fano effect is caused by the spin-orbit interaction of the photoelectron in the continuum. The explanation for it can be traced back to some effects of the same nature, namely the anomalous ratio of the doublet ^2P line-strength of alkali atoms (usually this ratio is two according to the statistical weight of the doublet states!); this was already dealt with by Fermi[25]. Also, as pointed out by Seaton[26], the non-zero minima of the photoionization cross sections of sodium, potassium, rubidium and cesium should result from spin-orbit interactions. It follows from Fano's detailed calculations that the Fano effect can be explained by the same type of spin-orbit interactions as in the two examples above. The relevant matrix elements, describing the dipole transitions of the photoionization from the discrete 6s ground state of Cs to εp continuum states are affected by the spin orbit coupling. Spin-orbit coupling is absent in the initial s state, nor does it affect the angular intergration of the dipole operator. The radial integral

$$R(\varepsilon, j') = \int_0^\infty P(\varepsilon p j'; r) r\, P(6s\ 1/2, r) dr$$

of the dipole transition matrix element does however depend on the total angular momentum j' of the continuum eigen states ($\varepsilon P_{1/2,\ 3/2}$) since the spin orbit coupling in the interior of the atom draws in the nodes of the wave function with j' = 1/2 to some extent and pushes out those for j' = 3/2. This difference in the radial dependence of the wave functions is caused by the spin-orbit potential $V_{s\ell} \propto \frac{1}{r} \frac{dV}{dr} (\underline{\ell} \cdot \underline{s})$ which represents a slight change of the total potential $V_{tot} = V_m + V_{s\ell}$ (V_m is the model potential for a continuum electron of a given atom; $V_{s\ell}$ is negative for states with $j = 1/2$, i.e. attractive, or $V_{s\ell}$ is positive for states with $j = 3/2$, i.e. repulsive)*.

There is another interesting effect associated with circularly polarized light and which was only recently discovered by Fox, Kogon and Robinson[28] at New York University. These experimentalists reported observations of the ionization of atomic cesium by simultaneous absorption of three ruby-laser quanta. Their results also indicated that circularly polarized ruby light produces ionization about twice as efficiently as does linearly polarized light. This effect for multiphoton ionization seems at first strange since the atoms used were unpolarized. As you know the rate for single photon ionization of atoms as independent of polarization of the incident light. Any difference in the single photon ionization rate by using either linearly or circularly polarized light would certainly violate the mirror invariance. Now, what is the explanation for the difference in the multiphoton ionization process between circularly and linearly polarized light?

Several theoreticians[29-33] worked almost simultaneously on a quantitative description, and I suppose that Dr. Lambropoulos might supplement something in particular in connection with the possibility of producing polarized electrons by multi-photon processes induced by circularly polarized light. Let me only refer to a

---

*Weisheit and Dalgarno[27] modified the expression for the dipole matrix element, of the photoionization, by accounting for alkali core polarizability in the field of the valance electron. The dipole transition operator $-e\underline{r}(1 - \alpha_d(\omega)/r^3)$ where $\alpha_d(\omega)$ is the dynamic dipole polarizability of the core at the transition frequency ω.

qualitative explanation which was already suggested by the discoverers of this interesting effect. The idea is that a two photon ionization process should proceed via an intermediate state, which has to be a P state if you allow dipole transitions and start with alkali atoms in the ground state. The intensity of the incident laser beam is supposed to be high enough that a second photon induces the ionization from the intermediate state into the continuum. Because of the selection rule $\Delta m = \pm 1$ associated with circularly-polarized light-induced transitions, the ionizing process from the intermediate P state only allows transition into D states in the continuum (d wave of the photoelectron!). In other words: the sequence of transitions of two-photon ionization processes with circularly polarized light is characterized by the transitions $S \rightarrow P \rightarrow D$. However, because of the selection rule $\Delta m = 0$ associated with transitions induced by linearly polarized light, the two photon ionization process branches into two possible sequences of transitions $S \rightarrow P \rightarrow D$ and $S \rightarrow P \rightarrow S$. In other words: the two-photon ionization process with linearly polarized light is a two-channel process which should be different in strength from the one-channel process with circularly polarized light.

Finally I have to admit that this short introduction into the present field of activity on spin and polarization in atomic process is by no means complete, but very fragmentary and biased by my way of selecting the topics. It would however be regrettable should I leave out a most beautiful example of application of electron spin polarization I recently learned of. An old problem in the physics of weak discharges was the question what is the mechanism for the production of electrons and ions which sustains the discharge. Without going much into the details for complicated production and loss mechanism of electrons in weak discharge, physicists often considered that the Penning ionization process should be amongst the predominant production process for electrons in weak discharges. In a Penning ionization process the excitation energy between colliding metastable atoms provides the ionization required. Recently Walters and co-workers[34] studied such processes in an ingeneous experiment with a weak helium discharge. It is known from optical absorption measurement of the $2^3S \rightarrow 2^3P$ and $2^1S \rightarrow 2^1P$ lines in weak helium discharges that the metastable 2^3S_1 and 2^1S_0 states are highly populated compared to other short lived excited states. In the experiment of Walters and co-workers the metastable triplet states were spin orientated by optical pumping with circularly polarized light ($2^3S_1 \rightarrow 2^3P \rightarrow 2^3S_1$). As is immediately obvious the spin polarization of metastable states can then be transferred to the free electrons resulting from the Penning reaction $He^M + He^M \rightarrow He(1^1S_0) + He^+ + e$ (He^M stands for metastable helium atoms). The free electrons in the discharge were extracted. (The extracted current depends, of course, upon the conditions like gas pressure, the intensity of the discharge and so on.) The electrons extracted from the active discharge were Mott analyzed, a polarization as high as 10% with a current of several μA was measured. When the discharge was terminated by removal of the excitation source, residual ions and electrons diffuse to the container walls where they recombine. However, production of new electrons through metastable-metastable reactions persists for a period comparable to the metastable diffusion time to the walls. Thus, in contrast to the active discharge, metastable-

metastable reactions are isolated as the only source of electrons during an afterglow period. Hence it is expected[35] and observed that the polarization of electrons, up to 40%, extracted during the afterglow period of a pulsed helium discharge, is larger than the electron polarization from the active discharge. So far this may suffice for the evidence of this beautiful example of applications with polarized electrons. It not only gave the proof that the above Penning reaction is the predominant source of ionization in weak electrical discharges in helium, but it also confirmed[36] that spin angular momentum is conserved in metastable-metastable reactions. Last but not least Walters and co-workers also tried to improve their method of optically pumped helium discharge as a promising intense source of polarized electrons.

At the end of my introduction I just would like to raise questions or problems which are certainly of interest in connection with the interdisciplinary nature of this conference, e.g. is there any analogy of the Fano effect in nuclear physics? I do not know, for example, what happens when you have the photo disintegration of the deuteron with circularly polarized γ radiation, or to what results two or multiphoto disintegration might lead, and what similarities are to be expected.

BEDERSON

Professor Kleinpoppen correctly characterized the differential spin exchange data for potassium as "typical" because it was in fact the only differential spin exchange data, until very recently. Now, A. Kasdan and T. M. Miller at New York University have obtained similar

data for sodium. The most important thing about these
measurements is to be able to compare them with appropriate theories, since the cross section measurements
with spin-analysis is far more sensitive to the choice of
atomic parameters, and to the nature of the approximations
made, than measurements obtained without spin-analysis.
The principal point of these experiments is that you
can affect a direct comparison with theory, whereas normal
unanalyzed experiments require suitable averaging over
various spin orientations, generally involving more than
one interaction, which means that a direct comparison with
theory is very difficult and certainly not unique. This
second example is of elastic spin-exchange measurement
for electron-sodium scattering. In such work we determine
the ratio of the elastic differential spin exchange to
the full differential cross section and use the independent measurements of the latter to normalize the
elastic differential spin exchange. Comparison of close-
coupling calculations and experiments is unusually
successful in the case of the alkalis, for a specific
reason: the alkalis are one-electron system, and are
highly polarizable. The oscillator strengths of the
first excited state, for all the alkalis except lithium,
are very large, so that convergence is unusually good in
a close-coupling expansion. As Professor Kleinpoppen has
said, the observables in the experiments relate to various
squares of the scattering amplitudes, and specifically,
in the one-electron case there are two scattering
amplitudes, say f (the "direct") and g (the "exchange")
amplitudes. Since these amplitudes are complex, and
since there is one arbitrary phase, there are three independent experimental observables that could be
measured in a one-electron system. In the past only a

single observable, namely the "full" differential cross section has been studied, as a function of energy and angle. In our experiments, at N. Y. U. using spin-analysis, two of the three observables have been measured for sodium and potassium.

In the case of inelastic scattering the situation is more complex. Instead of there being just f and g, there is an f and g for every magnetic substate of the excited state. So, for example in going from an s ground state to a p excited state one has a pair of direct and exchange amplitudes for each of the three magnetic substates ($M_\ell = \pm 1, 0$). Because of parity considerations, the excitation scattering amplitudes for $M_\ell = \pm 1$ are equal. Therefore in the excitation of one-electron atoms, s → p, there are four scattering amplitudes. Taking the one arbitrary phase into account, this means that there are seven observables, and therefore seven independent measurements which must be made to completely determine the scattering problem, in the non-relativistic limit, for the one-electron atom, for s → p collisions. We have done similar measurements to those described above, on the resonance excitation 4s → 4p in potassium, which have yielded two out of these seven quantities. [M. Goldstein, A. Kasdan and B. Bederson, Phys. Rev. 5, A660 (1972)]. I might say that close-coupling theory in this case is not as good as in the elastic scattering, as could be expected for a two-state expansion. Finally, let me just mention that the reason that we have been able to do these experiments and obtain such information out of them is, as is well known, that it is much easier to polarize a beam of neutral atoms than it is to produce a polarized beam of electrons. The reason for this is, of course, that the atom can be polarized in a Stern-Gerlach experiment, which is the basic technique for producing polarized atoms. In the case of electrons, the reason that polarized electrons are new today in physics is because of the experimental difficulties related to their production. New techniques for producing polarized electrons conveniently and relatively inexpensively are only just now beginning to reveal themselves.

Nevertheless, it is not likely that it will ever be as simple to obtain polarized electrons for low-energy collision experiments, as it is to obtain polarized atoms. My own feeling at the moment is that the scattering of electrons by mercury at several hundred eV is still the most likely candidate for a practical source of polarized electrons for <u>atomic</u> scattering experiments.

<u>KLEINPOPPEN</u>: Could you please say something about the currents one gets by scattering of electrons on mercury?

<u>BEDERSON</u>: The currents we get are of the order of 10^{-10} to 10^{-11} A at low energies, under optimum conditions. Professor Reichert is an expert on that, and I would ask him to answer that question by giving some numbers.

<u>REICHERT</u>: I agree with Professor Bederson that today scattering on mercury is the simplest method for producing polarized electrons at low energies. In our laboratories we get polarized electron currents typically of $10^{-9} - 10^{-8}$ A at energies of a few eV and with a degree of polarization around 30% by using this method.

<u>MACEK</u>:

Since the early work of Mott in the 1930's it has been recognized that particles can be polarized by collisions between initially unpolarized targets and projectiles, owing to coupling between spin and orbital angular momentum. Electrons, nucleons and even composite systems such as light nuclei have been polarized by this method. Owing to the weakness of the spin-orbit interaction in atoms and ions, they have not been spin polarized by collisions between initially unpolarized atoms. Dr. Shakeshaft and myself have recently pointed out how the relatively weak spin-orbit interaction may none-the-less provide the mechanism necessary to produce spin polarized hydrogen atoms by electron transfer collisions with ground state Xe.

We employ Massey's abdiabatic criteria to arrive at this conclusion. Figure M1 summarizes the argument.

$$H^+ + Xe \rightarrow H + Xe^+(^2P_{1/2})$$
$$\Delta E = .17 eV, \quad v_{max} \sim b\, \Delta E/\hbar$$
$$b \sim 5 \text{ a.u.} \quad K.E._{max} \sim 25 \text{ eV}$$

$$H^+ + Xe \rightarrow H + Xe^+(^2P_{3/2})$$
$$\Delta E = 1.5 \text{ eV} \quad K.E._{max} \sim 1.8 \text{ keV}$$

Fig. M1

The top reaction represents the charge transfer reaction leaving the Xe^+ ion in its first excited state. This requires that .17eV of translational energy be converted into internal energy. According to Massey's criteria this cross section will maximize at a velocity given by $\Delta E/\hbar$, where b is a dimension of the order of the atomic distances involved. We take it equal to 5 a.m. (atomic units). The cross section is then expected to maximize at K.E. ~ 25eV. The ΔE for the competing reaction, i.e. electron transfer leaving the Xe^+ ion in its ground state is 1.5 eV. According to Massey's criteria this cross section maximizes at 1.8 keV. Since the two levels of Xe^+ are split by the spin orbit coupling we see that the spin orbit forces, although weak, considerably influence the reaction. We thus expect that the hydrogen atom could emerge spin polarized.

We have made detailed numerical calculations based on the impact parameter approximation using an expansion in atomic eigenstates to calculate the spin polarization as a function of scattering angle for 15 eV protons incident on Xe. Our results are shown in Fig. M2. Note the strong diffraction minima, with the polarization averaged over $1°$ - $5°$ is approximately 70%. Our results show how spin-orbit effects can have a profound effect even in heavy ion collisions.

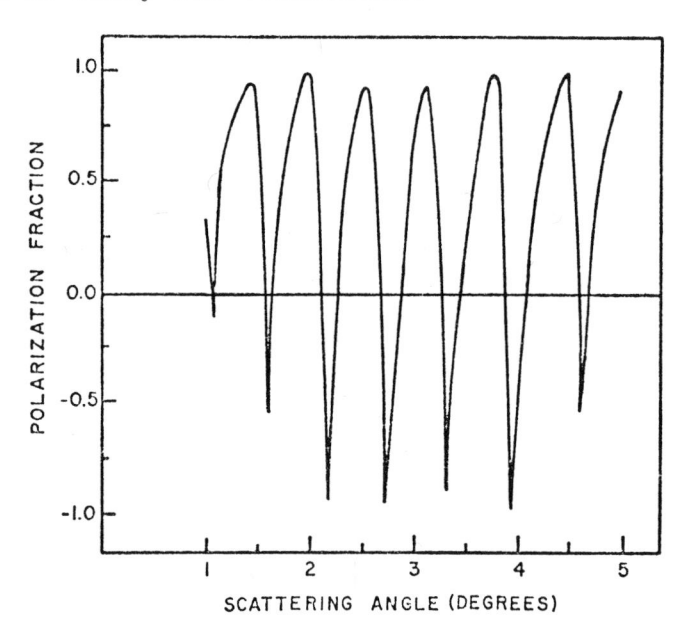

Fig. M2.

SPIN AND POLARIZATION EFFECTS

KLEINPOPPEN: What is the angular distribution of the scattered atoms?

MACEK: That is shown on Fig. M3. The important point to note is that the polarization maxima and the cross section maxima coincide.

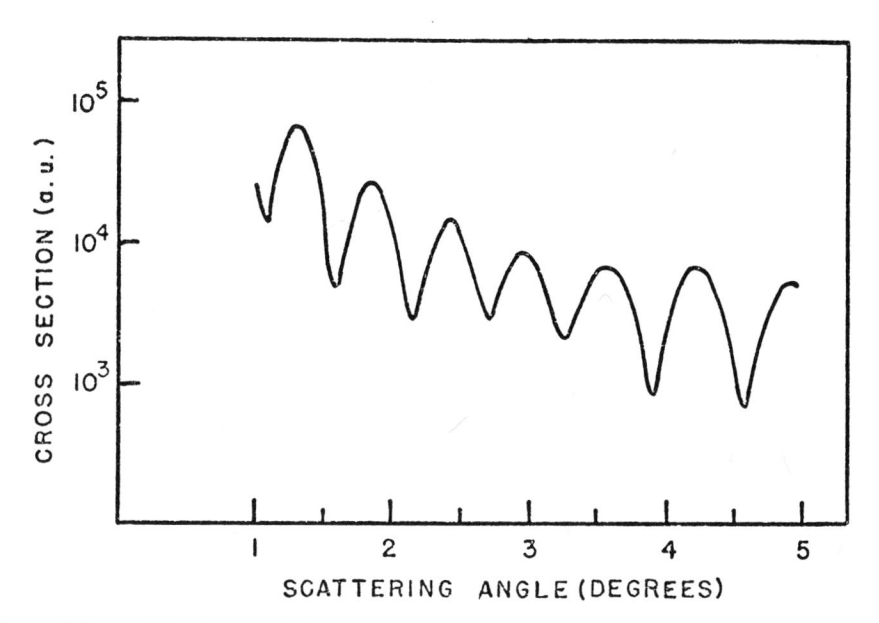

Fig. M3. Differential charge transfer cross section.

REICHERT: If you have a large angle of acceptance what is the mean polarization?

MACEK: The mean polarization is about 70% at 15 eV because the maximum of the spin polarization corresponds to the minima of the cross sections. This assumes you measure on one side of the zero degrees.

REICHERT: Would there be other systems where the same process occurs?

MACEK: It occurs in p-krypton scattering where we made calculations. But it should also occur in many other systems. The criteria that have to be satisfied are very simple.

LAMBROPOULOS (Electron Spin-Polarization Effects in Multiphoton Ionization.)

In the usual single-photon ionization of the alkali atoms with circularly polarized light,[22-24] the ejected photoelectrons have non-zero spin-polarization only when the radial matrix elements for the transition to the continuum obey the relationship

$$\rho \equiv <\epsilon P_{3/2}|r|nS_{1/2}> / <\epsilon P_{1/2}|r|nS_{1/2}> \neq 1;$$

i.e. when they depend on the total angular momentum. The photoelectron polarization attains the maximum value 1.0 when $\rho = -2$ which happens to occur near the minimum of the photoionization cross section as a function of photon frequency.

A similar phenomenon occurs in multiphoton ionization of the alkali atoms with circularly polarized light.[30] The resulting photoelectron spin-polarization in two- and three-photon near-resonance ionization is shown in Fig.L1. A basic difference between this case and single-photon ionization is that the polarization here is non-zero even when the radial matrix elements do not depend on the total angular momentum; i.e. even when all ratios ρ for bound-bound and bound-free transitions are equal to one. The only requirement is that the linewidth of the light source be smaller than the splitting of the near-resonant P-doublet; and of course that the light intensity be sufficiently high for the process to be observable. Both requirements are easily met with present day lasers.

In Fig. L1, it has been assumed that the photon frequency ω is approximately equal to the difference between the ground state and a P-doublet, and that it is tunable around the doublet. Thus the first transition is a near-resonant or resonant transition. Curves 1 and

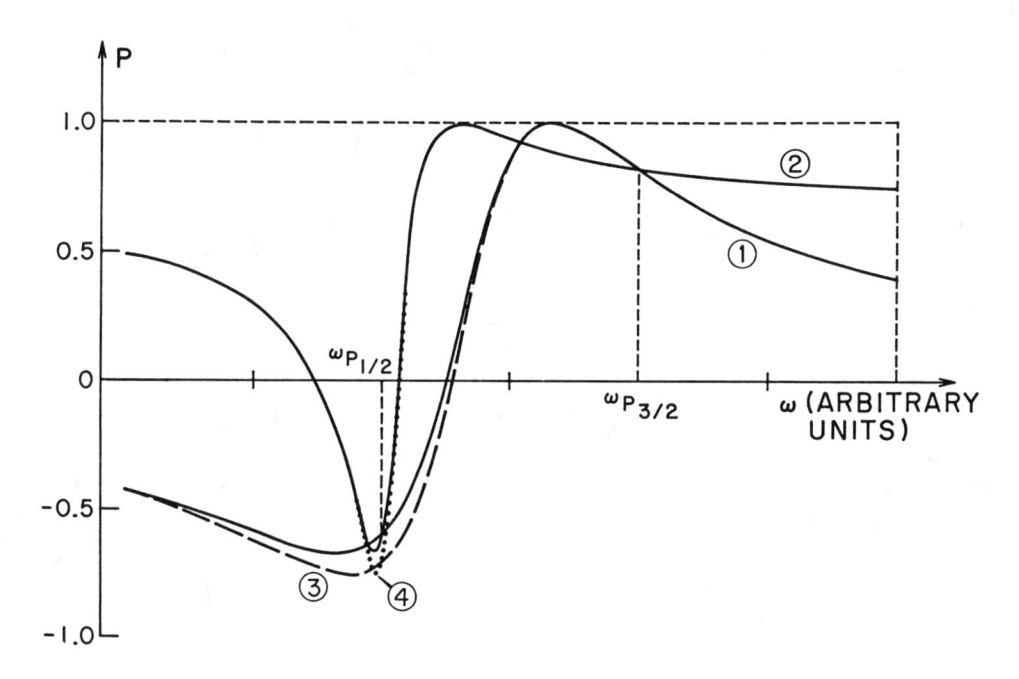

Fig. L1. Photoelectron polarization, P, for 2- and 3-photon ionization via a P near-resonance as a function of photon frequency, ω, in the vicinity of the P-levels. Curve 1: 2-photon ionization for

$$\rho \equiv <n'P_{3/2}|r|nS_{1/2}>/<n'P_{1/2}|r|nS_{1/2}> = 1.$$

Curve 2: 2-photon ionization for $\rho = 5$. Curve 3: 3-photon ionization for $\rho = 1$. Curve 4: 3-photon ionization for $\rho = 5$.

3 show the total (integrated over 4π) photoelectron spin-polarization in two- and three-photon ionization, respectively, for the case in which all ratios ρ are equal to one. This, for example, would be the case in Na. Curves 2 and 4 correspond to two- and three-photon ionization for the case in which

$$<n'P_{3/2}|r|nS_{1/2}>/<n'P_{1/2}|r|nS_{1/2}> = 5\quad,$$

where $n'P_j$ is the P-doublet in resonance with the photon

frequency. This, for example, would be the case in Cs for n' = 10 or 11.

It should be noted that as ω moves away from the doublet, the polarization goes to zero in cases $\underline{1}$ and $\underline{3}$, but it remains finite in cases $\underline{2}$ and $\underline{4}$. This results from the fact that in $\underline{1}$ and $\underline{3}$ the polarization is due to the level splitting alone. As ω moves too far away from the doublet, the splitting is no longer seen and the polarization drops to zero. In $\underline{2}$ and $\underline{4}$, however, the polarization is also due to the ratio ρ being different from one, and this contribution persists even far away from resonance. The exact asymptotic value depends on the value of ρ. Of course, if ω moves too far away from the doublet, other P-levels will become influential. In three-photon ionization, one can also have a second resonance, but this case will not be discussed here.

Unlike the single-photon case, the present effect occurs near resonances with intermediate states where the cross sections have maxima. It appears therefore that the effect can be used to produce highly polarized and relatively intense electron beams. Estimates indicate that by using present day lasers, one can obtain electron currents of the order of 10^{-5} A or more with polarization ~100%. Such beams would be pulsed with the pulse duration determined by the laser. This can vary from a few nanoseconds to milliseconds. Experiments in single-photon ionization of Cs have so far produced electron currents of the order of 10^{-11} A with polarization ~85%.

The above discussion assumes that the atoms themselves are unpolarized. When the atoms are polarized, several new effects occur which are of interest from a physical as well as practical standpoint.

GERJUOY: You showed these curves but to what angle of ejection do they refer?

LAMBROPOULOS: It is assumed that all the photoelectrons are collected. Polarizations of around 30% at 1μA.

KLEINPOPPEN: As far as I remember the Munich group obtained high polarization of about 80-90% from field-emission experiments.

Comment by P.S. FARAGO

In the previous discussion two scattering processes were mentioned which yield polarized electrons, Mott scattering and spin exchange collision. It is important to realize that the two processes are fundamentally different, although some phrases, such as "spin flip cross section" is commonly used in both contexts.

In Mott scattering the underlying mechanism is spin orbit coupling. By virtue of its motion in a Coulomb field the electron "sees" a magnetic field and the magnetic moment associated with the spin interacts with this field. In other words the force on the electron becomes explicitly spin dependent. The polarization of an initially unpolarized electron beam by Mott scattering arises by a selection process analogous to that observed in a Stern-Gerlach experiment.

Electron polarization by spin exchange scattering requires a spin polarized target (e.g. alkali atoms in one of the magnetic substates of the ground state) and the polarization of the scattered electrons is the result of the implications of the Pauli principle alone without the presence of explicitly spin dependent forces. From the Pauli principle it follows that the scattering of electrons from a polarized target can proceed along a singlet or along a triplet channel. The two processes have different amplitudes because the spatial part of the wave function has different symmetry properties in the two cases; one might say that in the two cases the electron moves in a different region in the vicinity of the scatterer. The proportion of "direct" to "spin-exchange" events depend on contributions from these two channels, and under favorable circumstances initially unpolarized electrons can become polarized at the expense of the depolarization of the target.

The fundamental difference between the effect of spin orbit coupling and spin-exchange can be illustrated in another way. If an initially polarized electron beam is scattered the direction of the polarization, in general, undergoes a rotation. If spin orbit coupling is responsible for the rotation of polarization it will follow the pattern of the rotation of optical polarization by optically active media. If the underlying mechanism is spin-exchange interaction the rotation of polarization will be analogous to the Faraday-rotation of optical polarization. The reason is that in the first case the axis of rotation is defined by the normal to the scattering plane (which reverses sign if the electron motion is reversed), in the second case the axis of rotation is defined by the polarization vector of the target, and this is independent of the geometry of the scattering.

REFERENCES

1. N.F. Mott, Proc.Roy.Soc. (London), A 124, 425, 1929, and A 135, 429, 1932.
2. H.S.W. Massey, and C.B.O. Mohr, Proc.Phys.Soc. (London), A 177, 341, 1941.
3. J. Schwinger, Phys.Rev. 69, 681 A, 1946.
4. M. Heusinkveld and G. Freier, Phys.Rev. 85, 80, 1952.
5. H. Faissner, Ergebn.Exakt.Naturwiss 32, 180 1959 (Review with references).
6. T. Heindorff, J. Höfft and E. Reichert, J.Phys.B to be published.
7. J.A. Simpson and U. Fano, Phys.Rev.Lett. 11, 158, 1963.
8. D.G. Thomson, J.Phys.B, 4, 468, 1971.
9. P.G. Burke, and J.F.B. Mitchell, J.Phys.B, to be published.
10. K. Blum and H. Kleinpoppen, Phys.Rev.A, to be published.
11. B. Bederson, J. Eislinger, K. Rubin and A. Salop, Rev.Sci.Instr. 31, No. 8, 1960.
12. K. Rubin, B. Bederson, M. Goldstein and R.E. Collins, Phys.Rev. 182, 201, 1969.
13. R.E. Collins, B. Bederson, and M. Goldstein, Phys. Rev.A 3, 1976, 1971.
14. E.M. Karule, Latvijas PSR Zinātnu Akadēmijas Vestis, p. 9, No. 3, 1970.
15. D.M. Campbell, H.M. Brash and P.S. Farago, Phys.Lett. 36A, 449, 1971.
16. E.M. Karule, and R.K. Peterkop, Atomic Collisions (ed. V. Ya.Veldre; Riga, USSR: Latvian Academy of Sciences) vol. 3, 1965.
17. D. Hils, M.V. McCusker, H. Kleinpoppen and S.J. Smith,

Phys.Rev.Lett. <u>29</u>, 398, 1972.
18. H. Kleinpoppen, and R. Neugart, Z.Physik <u>198</u>, 321, 1967.
19. D.R. Flower, and M.J. Seaton, Proc.Phys.Soc. <u>91</u>, 321, 1967.
20. H. Hafner, H. Kleinpoppen and H. Krüger, Phys.Lett. <u>18</u>, 270, 1965 and also H. Hafner, and H. Kleinpoppen, Z.Physik <u>198</u>, 315, 1967.
21. I.C. Percival and M.J. Seaton, Phil.Trans.Roy.Soc. London, A <u>254</u>, 133, 1958.
22. U. Fano, Phys.Rev. <u>178</u>, 131, 1969.
23. U. Heinzmann, J. Kessler, and J. Lorenz, Phys.Rev. Lett. <u>25</u>, 1325, 1970 and also Z.Physik <u>240</u>, 42, 1970.
24. G. Baum, M.S. Lubbell and W. Raith, Phys.Rev.Lett. <u>25</u>, 267, 1970.
25. E. Fermi, Z.Physik <u>59</u>, 680, 1930.
26. M.J. Seaton, Proc.Roy.Soc. A <u>208</u>, 418, 1951.
27. J.C. Weisheit and A. Dalgarno, Phys.Rev.Lett. <u>27</u>, 701, 1971.
28. R.A. Fox, R.M. Kogan, and E.J. Rubinson, Phys.Rev. Lett. <u>26</u>, 1416, 1971.
29. P. Lambropoulos, Phys.Lett. <u>40A</u>, 199, 1972, Phys.Rev. Lett. 29, 453, 1972.
30. P. Lambropoulos, Phys.Lett. <u>30</u>, 413, 1973.
31. S. Klarfeld, and A. Maquet, Phys.Rev.Lett. <u>29</u>, 79, 1972.
32. H.R. Reiss, Phys.Rev.Lett. <u>29</u>, 1129, 1972.
33. J. Mizuno, J.Phys.B, <u>6</u>, 314, 1973.
34. M.V. McCusker, L.L. Hatfield, and G.K. Walters, Phys.Rev.A, <u>5</u>, 177, 1972.
35. M.V. McCusker, private communication.
36. J.C. Hill, L.L. Hatfield, N.D. Stockwell and K.G. Walters, Phys.Rev. <u>A5</u>, 189, 1972.

ELECTRON-MOLECULE SCATTERING

A. Herzenberg

Yale University

New Haven, Connecticut

The amplitude for the scattering of an electron off a molecule may contain both direct and resonant parts.

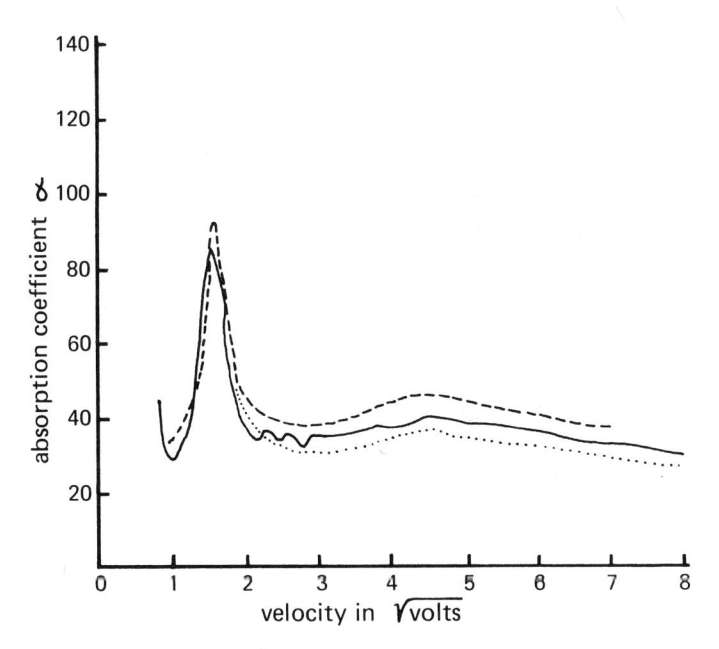

Figure 1 - The total cross-section of N_2 molecules for electrons, as it appeared in 1930. (See reference 1: Full curve, Normand; dashed curve, Brode; dotted curve Brüche.) Compare Figure 7 for more recent data.

TABLE I

IMPORTANT SHAPE RESONANCES IN DIATOMIC MOLECULES

line 1	Target molecule and vacant antibonding orbital	$H_2(\sigma_u)$	$N_2(\pi_g)$	$O_2(\pi_g)$
line 2	Minimum ℓ in the vacant antibonding orbital	$\ell \geq 1$	$\ell \geq 2$	$\ell \geq 2$
line 3	Rotational parameter $\dfrac{\Delta E_r}{\Gamma_n}$ (room temp.)	$\sim 10^{-2}$	$\sim 2 \times 10^{-2}$	$\gtrsim 5$
line 4	Vibrational parameter $\dfrac{\hbar\omega}{\Gamma_n}$	~ 0.2	~ 1	$\gtrsim 200$
line 5	Validity of the impulse approximation	Valid for rotations and vibrations	Valid for rotations but <u>not</u> vibrations	Impulse approximation useless
line 6	Resonance energy	~ 3 eV	~ 2 eV	0-1 eV

The first evidence for the existence of a resonance came about 1930 (Fig. 1) with a well-defined peak in the total cross-section for e-N_2 scattering at ~ 2 eV.[1] I shall restrict this talk to resonance scattering.

The nuclei move much more slowly than the electrons, so for the moment let us regard the nuclei as fixed. Then there can be Feshbach resonances in which an incoming electron becomes temporarily bound to an excited electronic state of the target. Such resonances lie

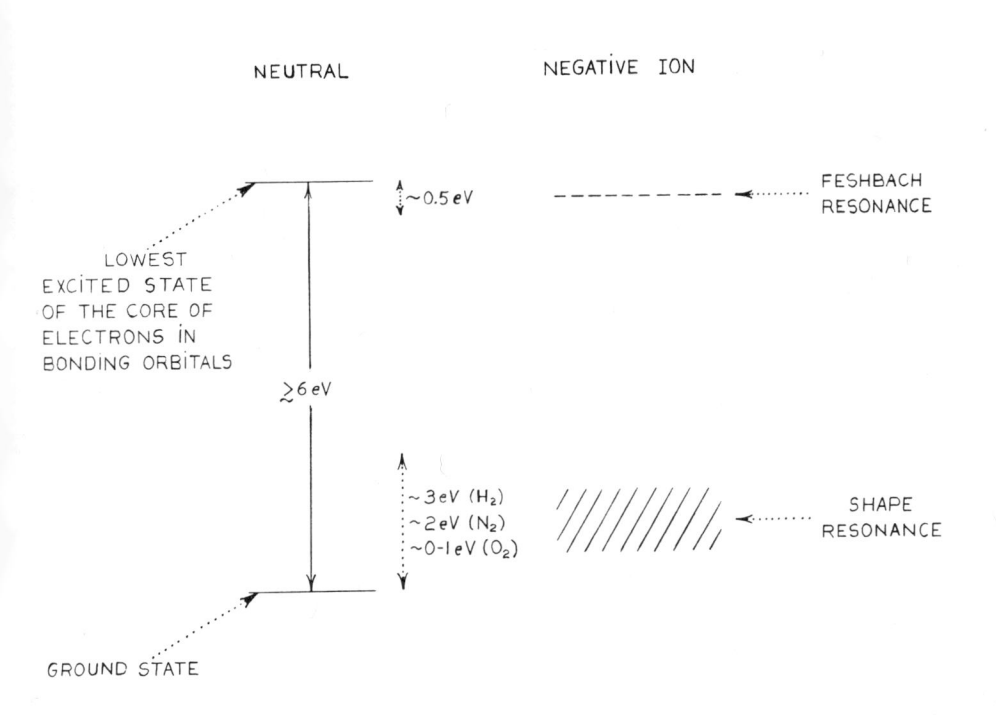

Figure 2 - Level scheme for a neutral diatomic molecule and its negative ion. The lowest electronic excited state of the core electrons in the neutral usually lies about 6 eV above the ground state, or higher. The electronic excited states of the neutrals sometimes give rise to Feshbach resonances, typically ~ 0.5 eV below the parent. Several diatomics also have shape resonances at much lower energy, due to the trapping of an electron within the centrifugal barrier of a vacant antibonding orbital.

typically from 6 eV upwards, because the lowest electronic excited state of a small target molecule possessing a set of fully occupied orbitals lies typically at 6 eV or higher, while the binding energy to the excited state is usually 0.5 eV or less (Fig. 2). However, there is often a shape resonance well below the lowest Feshbach resonance, that is a compound state formed by putting the incoming electron into a vacant orbital (see Figure 2). There are important examples of such resonances in H_2, N_2, and O_2; their salient properties are summarized in Table I. In each case, the bonding orbitals which can be constructed from the valence shells are full, and there are some vacant antibonding orbitals at a positive energy (see Figure 3). Such an antibonding orbital can trap an electron temporarily if there is a centrifugal barrier. For the existence of this barrier, the node between the nuclei in the antibonding orbital is very important; it raises the component of smallest angular momentum in a spherical harmonic exapnsion about the mid-point of the line joining the nuclei by at least one unit above the angular moments of the atomic valence orbitals from which it is constructed. For example, in H_2, the antibonding orbital is 1s-1s (Fig. 3), which has a minimum of $\ell = 1$ in a spherical harmonic expansion about the mid-point of the line joining the nuclei by at least one unit above the angular momenta of the atomic valence orbitals from which it is constructed. For example, in H_2, the antibonding orbital is 1s-1s (Fig. 3), which has a minimum of $\ell = 1$ in a spherical harmonic expansion about the mid-point of the molecular axis. In the same way, there are antibonding orbitals with $\underline{\ell \geq 2}$

ELECTRON-MOLECULAR SCATTERING   265

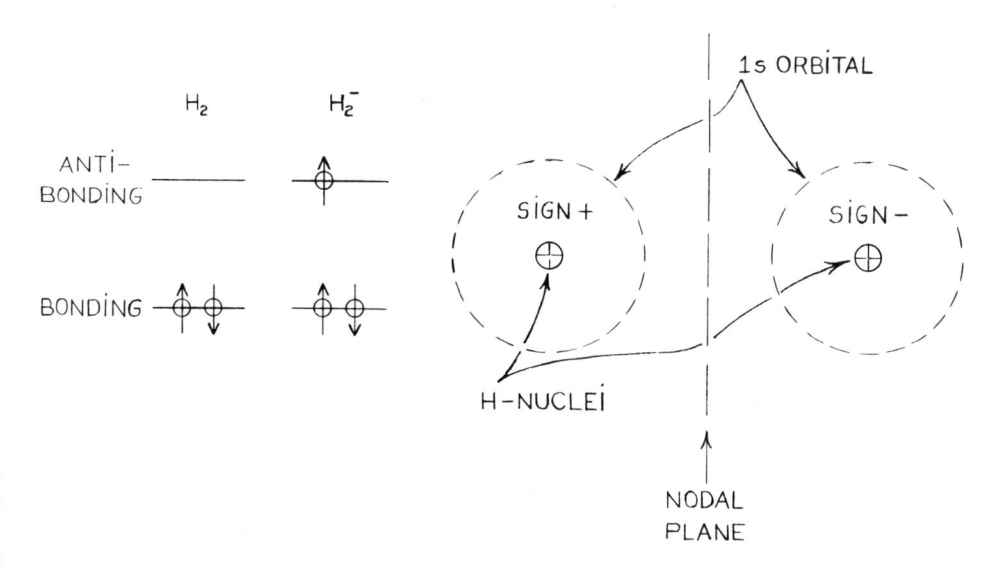

Figure 3 - The occupation scheme in the electronic ground state of H_2 and the lowest compound state of H_2^-. The diagram on the right shows the antibonding orbital in which the extra electron is trapped; the nodal plane leads to an asymptotic p-wave behaviour, and hence a centrifugal barrier.

in N_2 and O_2 which can be constructed from the 2p valence orbitals of the atoms. The minimum ℓ values are listed in line 2 of Table I. In each case -- H_2, N_2, and O_2 -- there is a low-lying shape resonance due to the occupation of the vacant antibonding orbital (at ~ 3 eV, ~ 2 eV, and 0-1 eV respectively).

Attempts to calculate the properties of these shape resonances have been made by several people[2,3]. One pretends that the nuclei are fixed, and calculates a complex energy

$$W_n(R) = E_n(R) - \frac{i}{2}\Gamma_n(R) \qquad \ldots \quad (1)$$

for each separation R of the nuclei (the suffix n distinguishes the resonances from one another); $\hbar/\Gamma_n$ is the lifetime of the corresponding compound state against emission of an electron. (If one could hold the nuclei fixed while one scattered the electron, the scattering amplitude would have a pole at $W_n(R)$). Results for the simplest case -- the lowest shape resonance in e + H_2 -- are shown in Figure 4 (from Reference 2). These curves were calculated by surrounding the molecule with a fictitious spherical box, solving the Schrödinger equation inside, and matching the wave-function to outgoing waves

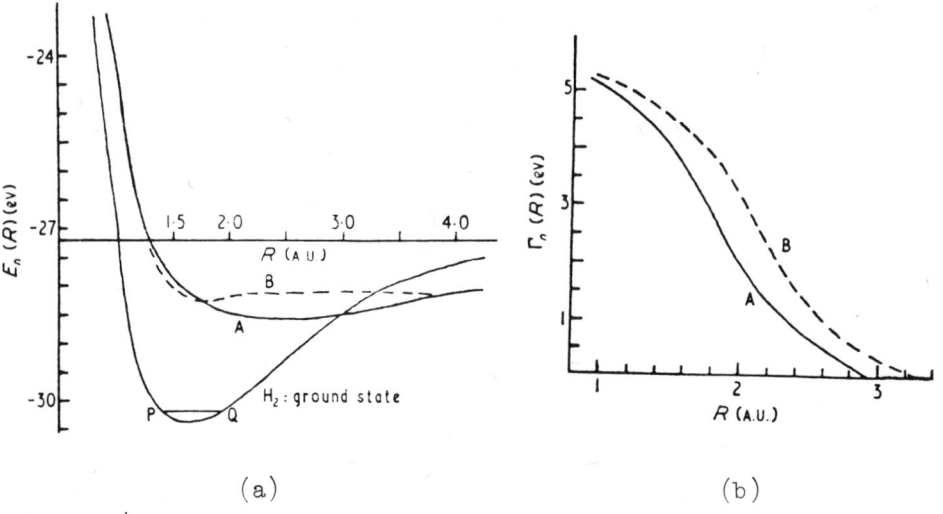

(a) (b)

Figure 4

(a) The real part of the energy of the $^2\Sigma_u^+$ shape resonance in H_2^- from the orbital in Figure 3; the full curve A and dashed curve B correspond to slightly different approximations. The H_2 ground state is calculated with a mixture of valence-bond and molecular orbital wave-functions constructed from hydrogenic 1s states.

(b) The imaginary part (x-2) of the complex energy of the $^2\Sigma_u^+$ shape resonance in H_2^-. The curves A and B correspond to A and B in Figure 4(a). From Reference 2.

at the surface. The full and dashed curves correspond
to slightly different approximations in carrying through
this procedure. The curves show that a "Franck-Condon"
range of resonance energies with a width of order ~ 1 eV
is swept through in the vibrational ground state of the
molecules. The decay width Γ_n of the compound states
varies from one compound state to another; in Figure 4
it is ~ 4 eV; in the state responsible for the peak in
Figure 1, it is about 0.6 eV[4]. The combination of Γ_n
and the Franck-Condon effect is responsible for the observed width of the peak in Figure 1; this large width
is responsible for the observation of an electron-molecule resonance nearly 40 years before the observation of
the first electron-atom resonance in 1964 (Schulz[5] saw
a resonance in e + He at 19.3 eV, with $\Gamma_n \approx 10^{-2}$ eV!).

The low lying shape resonances in electron-molecule
scattering are particularly simple because they involve
essentially only the motion of a single electron in a
potential well which changes slowly as the nuclei move.
There is of course some polarization of the target electrons, but this does not change the one-extra-electron
picture in a qualitative way. As a consequence of the
low energy, the shape resonances excite generally only
the vibrational and rotational states of the nuclei,
and not the electronic states.

Just how an incident electron is affected by a shape
resonance may be illustrated by a one-dimensional model
(Fig. 5), consisting of a potential hole (the interior
of the molecule) surrounded by a barrier*. For the sake
of simplicity, suppose the potential inside the barrier

---

*This is a schematic model for the centrifugal barrier,
modified by long range interactions; the precise shape
does not affect the qualitative conclusions.

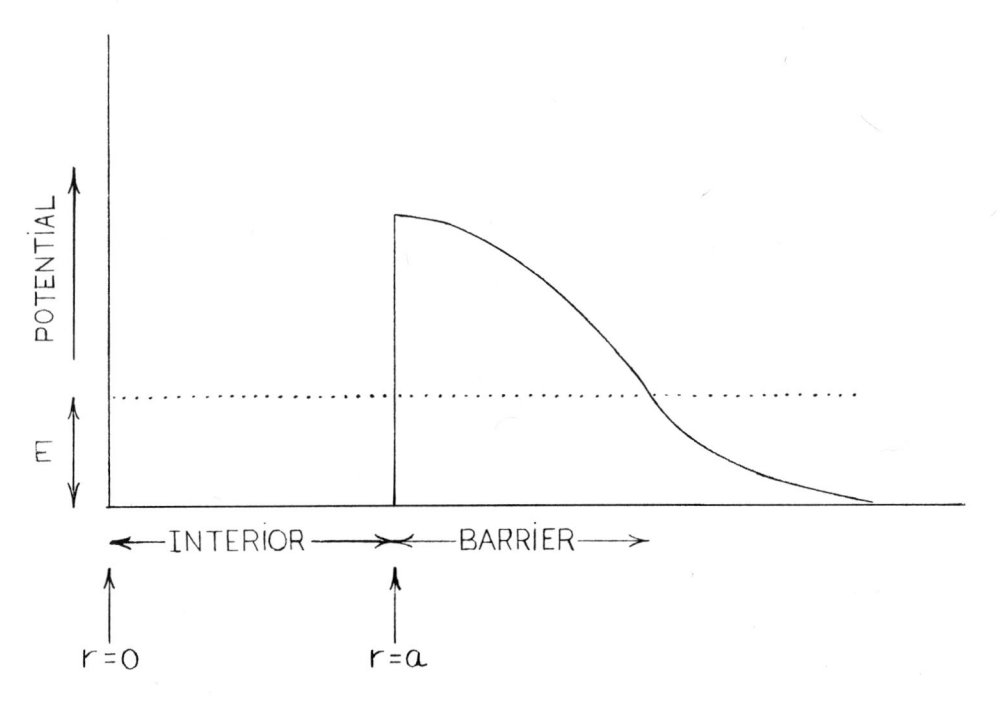

Figure 5 - A schematic one-dimensional model of a molecule surrounded by a centrifugal barrier. The precise shape of the barrier does not affect the qualitative conclusion in the text.

to be the same as outside. Suppose a wave e^{-ikr} (where k is the wave-number) impinges on the barrier from outside. The amplitude of the wave which penetrates into the well is Te^{-ikr}, where T is the amplitude transmission coefficient of the barrier. This wave is then perfectly reflected at the origin $r = 0$,* to yield an outgoing wave $- T e^{+ikr}$. There is another reflection at the barrier to yield an ingoing wave $(- R e^{2ika}) \times (- T e^{-ikr})$, where R is the amplitude reflection coefficient and where the factor e^{2ika} allows for the optical path within the well;

---
*The amplitude being discussed is r times the radial factor in the wavefunction, and therefore vanishes at $r = 0$.

ELECTRON-MOLECULAR SCATTERING

a is defined in Figure 5. This wave is again reflected perfectly at the origin, and so forth. If one adds up the initial penetrating wave $T\,e^{-ikr}$, and all the other components arising from multiple reflection, one gets

$$(e^{-ikr} - e^{+ikr})T\{1 + Re^{2ika} + (Re^{2ika})^2 + \ldots\}$$

$$= -2i\sin kr \left(\frac{T}{1 - Re^{2ika}}\right). \quad \ldots(2)$$

Let us suppose that the barrier is so strong that $|R| \approx 1$. Then the amplitude is generally of order T, and since $|T| \ll 1$ for a strong barrier, the amplitude inside the barrier is very much smaller than outside. An exception occurs at the special energies E_n where $R\,e^{2ik_n a} = |R|$ (k_n is related to E_n by $E_n = \hbar^2 k_n^2/2m$, where m is the mass of the projectile). The the amplitude inside becomes $T/(1 - |R|)$, and since $1 + |R| \approx 2$, and $|R|^2 + |T|^2 = 1$, one gets

$$\frac{T}{1 - |R|} \approx \frac{T}{\frac{1}{2}(1 - |R|^2)} = \frac{2}{T^*}. \quad \ldots(3)$$

Therefore if $|T| \ll 1$, the amplitude inside is now much greater than outside, essentially because all the multiply-reflected waves add up in phase; this is the condition of resonance. Outside the barrier, the amplitude of the scattered wave consists of two parts: first, a direct component with an amplitude of order one, which arises at the initial reflection outside the barrier; secondly, a component from the penetration through the barrier of the outgoing wave part inside the well, which is of order $T^*\{-T/(1 - R\, e^{2ika})\}$.[†] At most energies the second term is of order T^2 and therefore quite insignificant, but near resonance the second component becomes of order $|T|^2 \times (1/|T|^2) = 1$, and therefore comparable with the direct component. When $E \simeq E_n$, one can make a Taylor expansion of the denominator to get

$$\frac{-|T|^2}{1 - R\, e^{2ika}} \propto \frac{\Gamma_n}{E - E_n + (i/2)\Gamma_n} \qquad \ldots (4)$$

Γ_n comes out to be $\Gamma_n/\hbar \simeq (v/2a)|T|^2$, where v is the velocity; this is the product of the rate $(v/2a)$ at which the particle hits the barrier and the probability of penetration $|T|^2$. Here we have the Breit-Wigner formula for the amplitude.

The pole in the amplitude (4) at $E = E_n - \frac{i}{2}\Gamma_n$ leads to a time delay in the scattering of a projectile which was trapped after striking the target in the form of a wave-packet with average energy E_n. Suppose that the length of the wave packet is such that in the absence of trapping it would pass over the target in a time short compared with $\hbar/\Gamma_n$; then at a time t after the arrival of the wave-packet, there is left behind a part of the initial probability which decays with time as $\exp(-t\Gamma_n/\hbar)$; thus there is a time delay of order $\hbar/\Gamma_n$ before re-emission of the projectile.

The effects which are most significant for the excitation of a molecule through a resonance are the amplitude enhancement described by (3), and the time delay $\hbar/\Gamma_n$, which raises the time for which an electron can interact.

―――――
[†]If the amplitude penetration coefficient from right to left is T, the coefficient from left to right is T^*.

It is therefore not surprising that the three shape resonances are characterized by strong vibrational excitation, to which we shall return later. (Evidence on rotational excitation is more restricted due to the experimental difficulty of resolving individual rotational states).

A calculation of cross-sections has to recognize the difference between the electronic and nuclear velocities. This difference is characterized by the ratios

$$\frac{\text{Electronic delay time } (\approx \hbar/\Gamma_n)}{\text{Rotational period } (\approx \hbar/\Delta E_r)} \approx \frac{\Delta E_r}{\Gamma_n}, \quad \ldots(5)$$

and

$$\frac{\text{Electronic delay time } (\approx \hbar/\Gamma_n)}{\text{Vibrational period } (\approx \frac{1}{\omega})} \approx \frac{\hbar\omega}{\Gamma_n} \quad \ldots(6)$$

where ΔE_r is a typical rotational energy level spacing for the molecule, and $\hbar\omega$ a vibrational quantum. Γ_n is an average over the values of $\Gamma_n(R)$ in formula (1). The values of the ratios are listed in lines 3 and 4 of Table 1.

It follows that in the case of the low shape resonance in $e + H_2$, one has $\Delta E_r/\Gamma_n \ll 1$ and $\hbar\omega/\Gamma_n \ll 1$, so that both rotational and vibrational displacements are small during the collision. This suggests that one should neglect the kinetic energies of rotation and vibration during the encounter, even though they are of course essential to define the initial and final states of the molecule. An appropriate decomposition of the scattering $\hat{T}$ matrix* was given 20 years ago by Chew and Goldberger[6] in their treatment of the impulse approximation for scattering problems. They considered a Hamiltonian

$$H = K + U + V, \qquad \ldots (7)$$

where V is the interaction between target and projectile, and K + U defines the initial and final states; in the present context, $U = K_v + K_r$, where K_r is the kinetic energy of rotation and K_v the kinetic energy of vibration of the nuclei; K is the kinetic energy of the target electrons plus the potential energy within the H_2 target. If Φ_a and Φ_b denote the initial and final states, Chew and Goldberger showed that one may write for the matrix elements of $\hat{T}$

$$(\Phi_b|\hat{T}|\Phi_a) = (\Phi_b|V\{\omega^{(+)} + \frac{1}{E_a + i\eta - H}[U,\omega^{(+)}]\}|\Phi_a);$$

$$\ldots (8)$$

(Φ_a and Φ_b contain the electronic, vibrational, and rotational state of the molecule as well as the spin and momentum of the projectile). In (8)

$$E_a = \text{initial energy}$$
$$\eta = 0+,$$
$$\omega^{(+)} = 1 - i \int_{-\infty}^{0} dx\, e^{\eta x}\, e^{i(K+V)x}\, V\, e^{-iKx}.$$

---

*The notation $\hat{T}$ distinguishes the scattering matrix from the barrier transmission T.

The significance of $\omega^{(+)}$ is that, operating on Φ_a, it gives the state which would arise at time = 0 if the state at time = $-\infty$ had been Φ_a, and U had been neglected. The first term in equation (8), i.e.

$$(\Phi_b | V \omega^{(+)} | \Phi_a) \qquad \ldots (9)$$

is called the impulse approximation; note that the operator $V\omega^{(+)}$ ignores the vibrational and rotational nuclear kinetic energies U.

To evaluate (9), it is convenient to write

$$\Phi_a(\underline{r},\underline{R}) = \underline{k}_a(\underline{r}) \, \psi_a(\underline{r},\underline{R}) \, \chi_a(\underline{R}),$$

$$\Phi_b(\underline{r},\underline{R}) = \underline{k}_b(\underline{r}) \, \psi_b(\underline{r},\underline{R}) \, \chi_b(\underline{R}).$$

where $\underline{r}$ stands for the electronic and $\underline{R}$ for the nuclear co-ordinates; $\underline{k}_a$ and $\underline{k}_b$ are the initial and final states of the projectile; ψ_a and ψ_b the initial and final states of the target electrons; χ_a and χ_b are the initial and final states of vibration and rotation of the nuclei. Let $t_{ba}(\underline{R})$ denote the scattering amplitude one would have if the nuclei were fixed, i.e.

$$t_{ba}(\underline{R}) \equiv \int d\underline{r} \, \underline{k}_b^*(\underline{r}) \, \psi_b^*(\underline{r},\underline{R}) \, V \omega^{(+)} \, \underline{k}_a(\underline{r}) \, \psi_a(\underline{r},\underline{R}).$$

The expression (9) is

$$(\Phi_b | V\omega^{(+)} | \Phi_a) = (\chi_b | t_{ba} | \chi_a) \equiv \int d\underline{R} \, \chi_b^*(\underline{R}) \, t_{ba}(\underline{R}) \, \chi_a(\underline{R});$$

$$\ldots (10)$$

in other words, the impulse approximation tells one to calculate the scattering amplitude as if the nuclei were fixed at $\underline{R}$, and then to evaluate a matrix element with the states χ_a and χ_b expressed as functions of $\underline{R}$.

The order of magnitude of the correction term in (8) may be estimated. For this purpose, it is convenient to add to U the potential energy $E_v(\underline{R})$ of vibration in the initial state of the bare molecule; this is permissible because $[E_v(\underline{R}), \omega^{(+)}] = 0$, because both are diagonal in the nuclear co-ordinates $\underline{R}$. Therefore

$$[U,\omega^{(+)}] = [K_v + E_v, \omega^{(+)}] + [K_r, \omega^{(+)}]. \quad \ldots (11)$$

Taking matrix elements with respect to vibrational and rotational states $|vJ\rangle$ of the bare molecule, one has

$$\langle vJ | [U,\omega^{(+)}] | v'J'\rangle =$$

$$[(E_v(v) - E_v(v')) + (E_r(J) - E_r(J'))] \langle vJ | \omega^{(+)} | v'J'\rangle ,$$

where $E_v(v)$ is the vibrational and $E_r(J)$ the rotational energy of state $|vJ\rangle$. Since near a broad resonance $\omega^{(+)}$ varies slowly with $\underline{R}$ in comparison with the nuclear wavelengths, one has $|E_v(v) - E_v(v')| \approx \hbar\omega$, and $|E_r(J) - E_r(J')| \approx \Delta E_r$, for important matrix elements. In the neighborhood of a resonance, the factor $1/(E_a + i\eta - H)$ in (8) is dominated by a resonance term with denominator $\frac{1}{2}\Gamma_n$. Therefore the correction term in (8) has vibrational and rotational contributions of order $\hbar\omega/\Gamma_n$ and $\Delta E_r/\Gamma_n$ relative to the leading impulse approximation term, or

$$(\Phi_b | \hat{T} | \Phi_a) = (\chi_b | t_{ba} | \chi_a) [1 + 0(\frac{\hbar\omega}{\Gamma_n}, \frac{\Delta E_r}{\Gamma_n})]. \quad \ldots (12)$$

It may happen that the molecular period of rotation $\hbar/\Delta E_r$ is large compared with the residence time $\hbar/\Gamma_n$ of the electron (i.e. $\frac{\Delta E_r}{\Gamma_n} \ll 1$), but that the vibration period $1/\omega$ is not, i.e. $\hbar\omega/\Gamma_n \sim 1$ or $\hbar\omega/\Gamma_n \gg 1$. For example, in the case of $e + N_2$ described in Table I, one has $\Delta E_r/\Gamma_n \approx .02 \ll 1$, $\hbar\omega/\Gamma_n \approx 1$.* In such a case, one may apply the impulse approximation to the rotations only; i.e. one sets $U = K_r$ (instead of $U = K_v + K_r$). Then in place of the former $V\omega^{(+)}$, which corresponded to fixed nuclei, one gets the scattering matrix for a fixed direction of the molecular axis, but with the vibration of the nuclei along that axis taken into account; we shall denote this operator by $V\omega^{(+f.a.)}$, with "f.a." standing for "fixed axis". Let us factorize the nuclear wavefunctions χ_a and χ_b in (10) into vibrational and rotational parts:

$$\chi_a(\underline{R}) = \chi_a^{(v)}(R) \chi_a^{(r)}(\hat{\underline{R}}),$$

$$\chi_b(\underline{R}) = \chi_b^{(v)}(R) \chi_b^{(r)}(\hat{\underline{R}});$$

correspondingly, let us define a new scattering amplitude $t_{ba}^{(f.a.)}(\hat{\underline{R}})$, which includes the vibration of the nuclei along their fixed axis, but not the rotation:

$$t_{ba}^{(f.a.)}(\hat{\underline{R}}) \equiv \int d\underline{r}\, d\underline{R}\, k_b^*(\underline{r})\, \psi_b^*(\underline{r},\underline{R})\, \chi_b^{(v)*}(R)$$

$$\times\, V\omega^{(+f.a.)}\, \underline{k}_a(\underline{r})\psi_a(\underline{r},\underline{R})\chi_a^{(v)}(R) \qquad \ldots (13)$$

Thus if one applies the impulse approximation to the

---

*Since Γ_n varies with the separation of the nuclei, one has to take a suitable average here. The value we have used corresponds to early calculations in which Γ_n was treated as a constant;[7] one found $\Gamma_n = 0.20$ V for a reasonable fit of experimental cross-sections, so that $\hbar\omega/\Gamma_n = 0.3 eV/0.2 eV - 1.5$).

rotations alone, the leading term in the complete scattering amplitude becomes

$$(\Phi_b|T|\Phi_a) = \int d\hat{\underline{R}}\, \chi_b^{(r)*}(\hat{\underline{R}})\, t_{ba}^{(f.a.)}(\hat{\underline{R}})\, \chi_a^{(r)}(\hat{\underline{R}}). \qquad \ldots (14)$$

The correction term in (8) now contains only the second term on the right in equation (11), so that its order of magnitude is smaller than the leading term (12) by a factor $(\Delta E_r/\Gamma_n)$.

If neither of the conditions $\hbar\omega/\Gamma_n \ll 1$ and $\Delta E_r/\Gamma_n \ll 1$ is satisfied, the decomposition in equation (8) ceases to be useful. This is the case for O_2 described in Table I, where other methods have to be used; this case has still not been analyzed.[8]

The simplest (and first) application of the impulse approximation in electron molecule scattering[9] at a resonance was to the rotational transitions in the collision e + H_2 at energies below 10 eV. Although the vibrations may be treated by the impulse approximation in this case, it is still convenient to write the amplitude for rotational excitation in the form of equation (14). Moreover, one may assume that vibrational excitation in this energy region ($\leq$ 10 eV) proceeds almost entirely through resonance scattering, because an early calculation by Carson[10] had shown that the Born approximation gives the cross-section for vibrational excitation far too small, whereas a resonance model[11] gives the right order of magnitude and energy dependence. The dependence of $t_{ba}^{(f.a.)}(\hat{\underline{R}})$ in 14 on $\hat{\underline{k}}_a$ and $\hat{\underline{k}}_b$, the directions of the incoming and outgoing electron, is very simple to write down when the resonance is due to the antibonding orbital in Figure 3; if one takes just the

ELECTRON-MOLECULAR SCATTERING 277

p-wave component into account, $\hat{\underline{k}}_a$, $\hat{\underline{k}}_b$, and $\hat{\underline{R}}$ enter through a factor $t_{ba}^{(f.a.)}(\hat{\underline{R}}) \propto \hat{\underline{k}}_a \cdot \underline{R} \; \hat{\underline{k}}_b \cdot \underline{R}$. The integral (14) then accounts fairly well for the excitation cross-sections and angular distributions of the rotational states accompanying the excitation of the first vibrational quantum. As an example, Figure 6 shows the ratio of the cross-sections for the excitation of one vibrational quantum

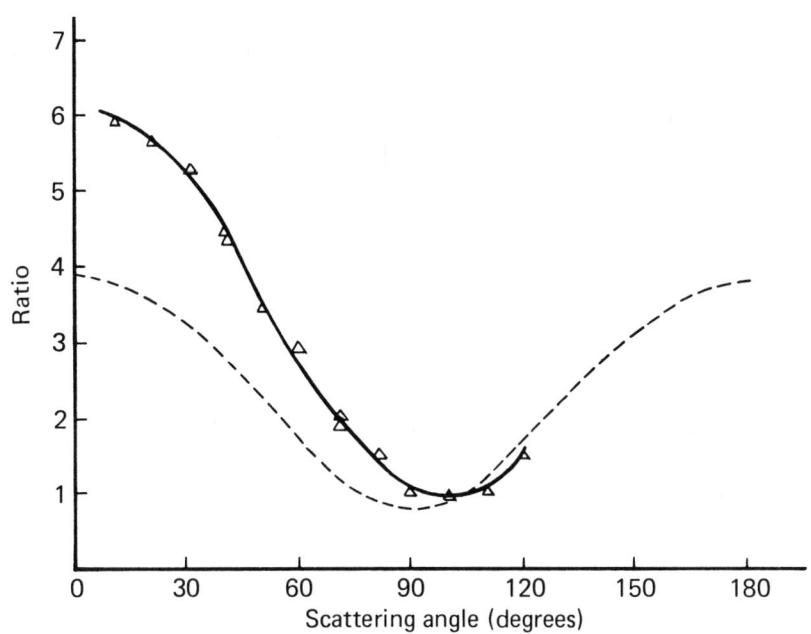

Figure 6

The ratio
$$\frac{I(\underline{k},\underline{k}'; \; v=0 \to v'=1; \; J=J')}{p_1 \; \sigma(\underline{k},v=0, \; J=1 \to \underline{k}',v'=1, \; J'=3)}$$

[I = cross-section for the excitation of one vibrational quantum without change of total angular momentum;
p_1 = fraction of molecules in the state $v=0$, $J=1$ at room temperature;
σ (in the denominator) = cross-section for the excitation of one vibrational quantum with change of total angular momentum from $J=1$ to $J'=3$].

Full curve - experiment, reference 12.

Dashed curve - impulse approximation, reference 9.

without change of the total angular momentum, and with the transition $J = 1 \to J = 3$ [9,12]. A discussion of the vibrational excitation by the impulse approximation has recently been given by Faisal and Temkin.[13]

The intermediate case $\hbar\omega \sim \Gamma_n$, which occurs in e + N_2 in Table I, is particularly interesting. Here the delay time is sufficiently large to permit a substantial displacement of the nuclei, and yet not large enough to permit a well-developed vibrational spectrum. However, it is still possible to apply the impulse approximation to the rotations, so that (13) and (14) may be used. In the integrand in (12), one should note that the factor $\omega^{(+f.a.)}{}_{\underline{k}_a}(\underline{r}) \psi_a(\underline{r},\underline{R}) \chi_a^{(v)}(R)$ represents the wavefunction when the extra electron is close to the molecule, since the limited range of the potential V permits only such configurations to count. If we assume that the wavefunction in this region is dominated by a resonant state $\psi_n(\underline{r},\underline{R})$ of the electrons, one must have

$$\omega^{(+f.a.)}{}_{\underline{k}_a}(\underline{r}) \psi_a(\underline{r},\underline{R}) \chi_a^{(v)}(R) = \psi_n(\underline{r},\underline{R}) \xi_n(R);$$

ξ_n is a function of the nuclear co-ordinates only, and varies with energy. If Γ_n had been small compared with $\hbar\omega$, one would have expected ξ_n to be large only if the energy coincided with one of the vibrational energies of the negative ion, and to be vanishingly small otherwise. A detailed discussion[4] shows that when Γ_n and $\hbar\omega$ are of the same order, ξ_n is quite different; there are no discrete vibrational states, and ξ_n is non-zero over a continuous band of energies about 1.5 eV wide in N_2. The main effect of a variation of the energy is to make the nodes and antinodes drift as the energy changes. This drift leads to oscillations in the cross-sections, because according to (13)

$$t_{ba}^{(f.a.)}(\hat{\underline{R}}) =$$

$$\int dR \, \chi_b^{(v)*}(R) \, \xi_n(\underline{R}) \, \{\int d\underline{r} \, k_b^*(\underline{r}) \, \psi_b^*(\underline{r},\underline{R}) V \, \psi_n(\underline{r},\underline{R})\},$$

where the integral { } contains only electronic wavefunctions and so must vary slowly with R in comparison with the rapidly oscillating factor $\chi_b^{(v)*}(R) \, \xi_n(\underline{R})$; as the nodes and antinodes of ξ_n drift, they pull across the oscillations in $\chi_b^{(v)}$ and so make the scattering amplitude oscillate.

Before the theory can be compared with experiment, one has to do the integral (14), take the squared modulus to get the cross-section, and sum over final and average over initial rotational states. This sum and average is essential because at present it is not possible to resolve individual final rotational states, or to start with the target in a definite initial rotational state. If one ignores the slight differences in the velocities of the outgoing electrons from the different rotational states, then the completeness of the rotational states make the summed and averaged cross-section to a particular vibrational state proportional to

$$\int d\hat{\underline{R}} \, |t_{ba}^{(f.a.)}(\hat{\underline{R}})|^2.$$

This is just the cross-section one would have if the axis were fixed in each collision, but oriented at random in different collisions.

The theory may be fitted to experiment by treating the electronic energy $W_n(R)$ of the negative ion parametically. The result for e + N_2 in Table I is shown in

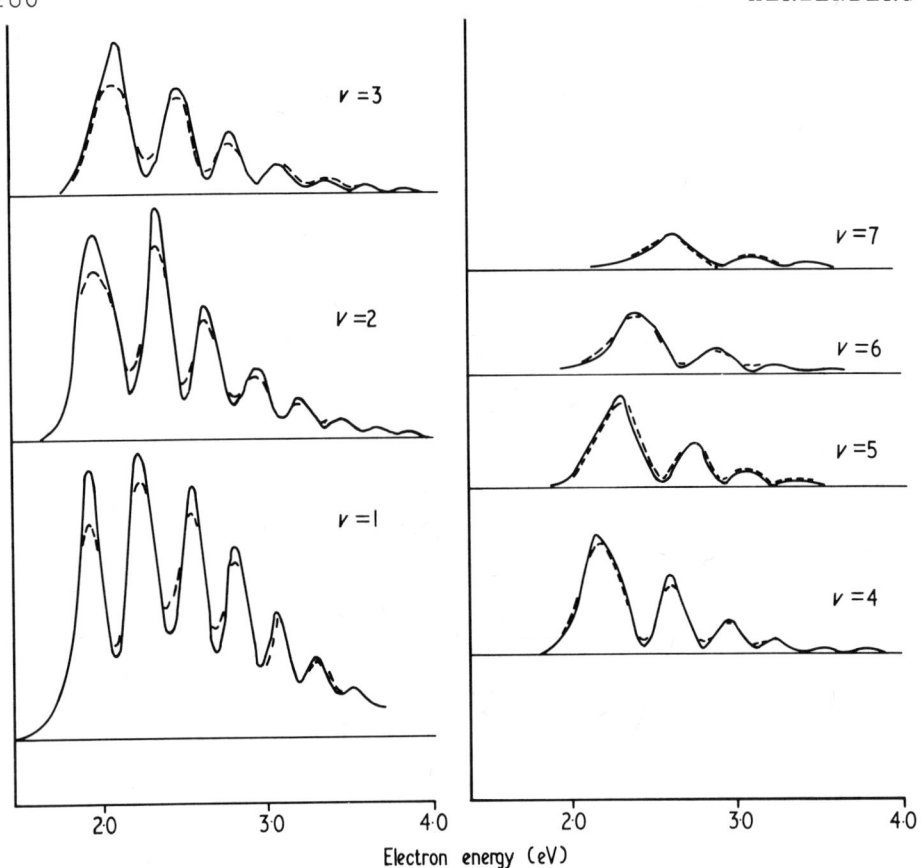

Figure 7 - Comparison of experimental (full curve) and calculated (dashed curve) cross-sections for vibrational excitation of the electronic ground state in N_2. (Arbitrary units; only relative cross-sections are significant. The experimental curves are due to Ehrhardt and Willmann, published in 1967; the oscillations were first observed by Schulz in 1962.) From reference 4.

Figure 7; the real part $E_n(R)$ was treated as an oscillator potential with anharmonic terms, while the imaginary part $-\frac{i}{2}\Gamma_n(R)$ was chosen in accordance with the d-wave centrifugal barrier through which the extra electron has to tunnel at entry and re-emission (Fig. 8). The values of the parameters used, and the calculational details are given in Reference 4. It turns out that the potential parameters one needs are in good agreement with an ab

ELECTRON-MOLECULAR SCATTERING

initio calculation by Krauss and Mies.³ The physical picture which emerges is illustrated in Figure 8; the key to understanding the peculiar shift of the maxima in Figure 7 from channel to channel turns out to be the idea that the strong maximum of $\Gamma_n(R)$ to the left in Figure 8 (where tunnelling is easiest) permits essentially only a

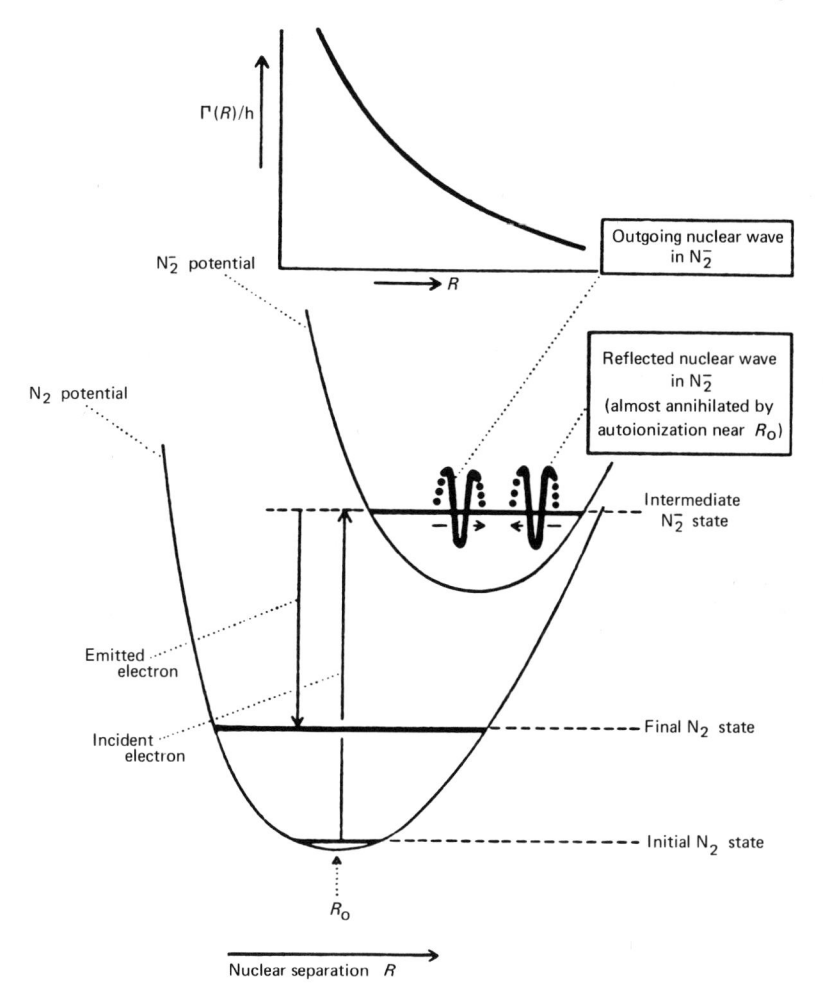

Figure 8 - The "boomerang" model of the nuclear wavefunction in the N_2^- ion, which suggested the calculation with variable Γ that led to Figure (7). The model was based on the assumption that the magnitude and R-dependence was such that only a single outgoing and a single reflected wave matter. (From reference 4).

single outgoing and returning nuclear wave in the negative ion; after that, the ion is destroyed by autoionization. (The details of the argument are given in Reference 14).

The appearance of the cross-section curves for vibrational excitation in the three cases in Table I is very different. That the theory can accommodate the different possibilities is demonstrated in Figure 9, which shows what happens when Γ_n is increased by a factor 4 or decreased by a factor 60 from the value required to give the fit in Figure 7. The enlarged Γ_n, in Figure 9(a), shows no oscillations, as in the case of e + H_2 in Table I. The diminished Γ_n, in Figure 9(b), shows a well-developed spectrum of vibrational states, as is observed in the case of e + O_2.[8]

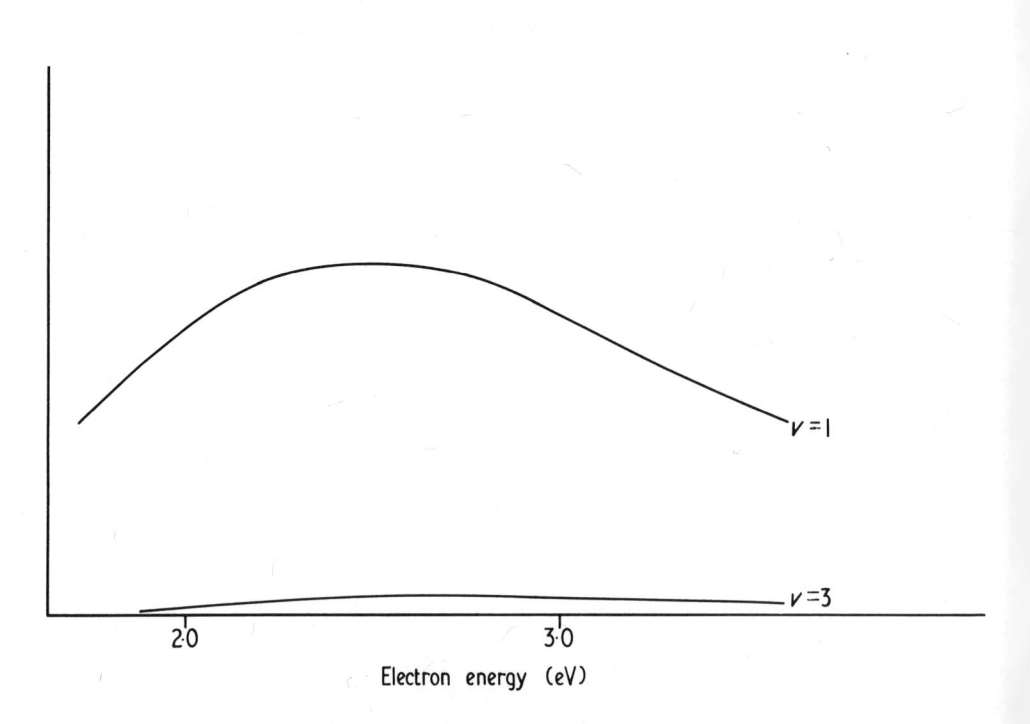

Figure 9a

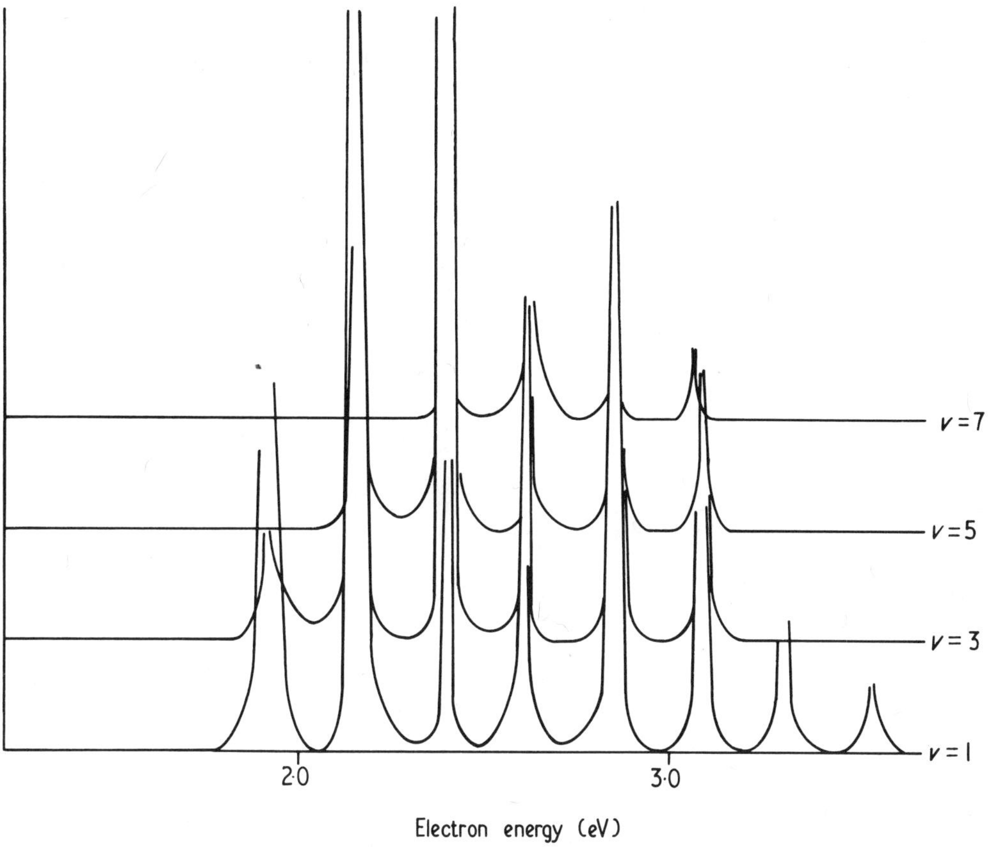

Figure 9b

Figure 9 - The change from the vibrational excitation cross-section shown in Figure 7 when Γ_n is increased by a factor 4 (see (a)) or decreased by a factor 60 (see (b)). The shape of curve (a) resembles the vibrational excitation cross-section due to the H_2 shape resonance in Table I the curve (b), with its discrete vibrational levels occurring at the same set of energies in each channel, resembles the observed cross-section for the shape resonance in O_2 (see reference 8). (From reference 4).

REFERENCES

1. Normand, C. E., Phys.Rev. 35 (1930), 1217
   Brode, R. B. Phys.Rev. 25 (1925), 636
   Brüche, E., Ann.Physik, 82 (1927), 912; 83, (1927), 1066
2. Bardsley, J. N. Herzenberg, A., and Mandl, F., Proc. Phys.Soc. 89 (1966), 305-19.
3. Krauss, M. and Mies, F. H., Phys.Rev. A 1 (1970) 1592-8.
4. Birtwistle, D. T., and Herzenberg, A., J.Phys.B (Atom. Molec. Physics) 4 (1971), 53
5. Schulz, G. J. Phys.Rev. Lett. 10 (1963), 104-5.
6. Chew, G. F. and Goldberger, M. L., Phys.Rev. 87 (1952), 778-82.
7. Herzenberg, A., and Mandl, F., Proc.Roy.Soc.A, 270 (1962), 48-71.
8. Linder, F., and Schmidt, H., Z. Naturforsch. 26a (1971), 1617-25.
9. Abram, R. A., and Herzenberg, A., Chem.Phys.Lett. 3 (1969), 187
10. Carson, T. R., Proc.Phys.Soc.A, 67 (1954), 908-16.
11. Bardsley, J. N., Herzenberg, A., and Mandl, F., Proc. Phys.Soc. 89 (1966), 321-40.
12. Ehrhardt, H., and Linder, F., Phys.Rev.Letters, 21 (1968), 419-21.
13. Faisal, F. H. M., and Temkin, A., Phys.Rev.Lett. 28 (1972), 203-6.
14. Herzenberg, A., J.Phys.B 1 (1968), 548-58.

PANEL DISCUSSION

ELECTRON MOLECULE COLLISIONS

        Panel Chairman: A. Temkin
                              (Goddard Space Flight Center)
        Panel Members: L. Frommhold (University of Texas), R.J. Henry (Louisiana State University), A. Herzenberg (Yale University), H.S.W. Massey (University College, London).

FROMMHOLD (Rotational Resonances)

Let me point out why I think there is some strong experimental evidence concerning the existence of a new resonance in the scattering of low-energy electrons by molecules. We call these resonances rotational resonances, because they can be viewed as a virtual excitation of a rotational transition. In this resonance the kinetic energy of the electron is below threshold for rotational excitation. Excitation is virtually possible, however, if the electron is temporarily attached to the molecule thus providing the required energy defect from its "affinity."

The evidence to be presented is not obtained from electron beam experiments, because in the past it proved to be impossible to do an experiment at a beam energy below the threshold for rotational excitation, i.e. at energies of ten or twenty millivolts. However, if an experiment involving a drift-tube is done measuring mobilities of electrons in a molecular gas, one can easily control the electron mean energies between about 6 and 1000 mV. If the drift velocities of electrons are measured under such conditions, an unexpected delay ("trapping") of the electron drift was observed at certain, fixed values of the mean electron energy. With increasing pressure the total delay was found to increase. This effect was first observed by J. J. Lowke[1]. Measurements of this "pressure effect" were later extended by several authors, mostly notably by R. Grünberg[2] and R. W. Crompton et al.[3] We were able to show that these experimental data were all consistent with the assumption that temporary electron trapping via resonance states takes place.

In Fig. 1 we are looking at Grünberg's data. Plotted are the reciprocal drift velocities as function of pressure at a constant ratio of fieldstrength over pressure, i.e. at a constant mean energy of the electrons. We plot the <u>reciprocal</u> drift velocities, which correspond to the transit time of the electrons per unit drift distance. Current theories of electron mobilities do not account for electron trapping and therefore predict zero slopes. Obviously the experimental data show non-zero slopes at least at some energies, which fact points out the need to account for trapping states.

Such a theory accounting for electron trapping can readily be developed[4], with the result that the electron transit times per unit length become a linear function of pressure:

$$\frac{1}{v_e} = \frac{1}{v_{oe}} (1 + \nu\tau)$$

Herein v_e and v_{eo} are electron drift velocities at the nominal and a very small pressure, respectively. ν is

ELECTRON MOLECULE COLLISIONS 287

Fig. 1. A selection of experimental reciprocal drift velocities, v_{e0}/v_e, versus the hydrogen pressure. As in this graph, a linear dependence is accurately observed for all the existing data.3-5 (No linear dependence of v_e versus p could be obtained at the larger density ranges of Ref. 2.) Their slopes α are plotted in Fig. 2, as function of the characteristic electron energies and gas temperature, and will be called $(\nu\tau)_l$ for reasons explained below. (Data from Ref. 2).

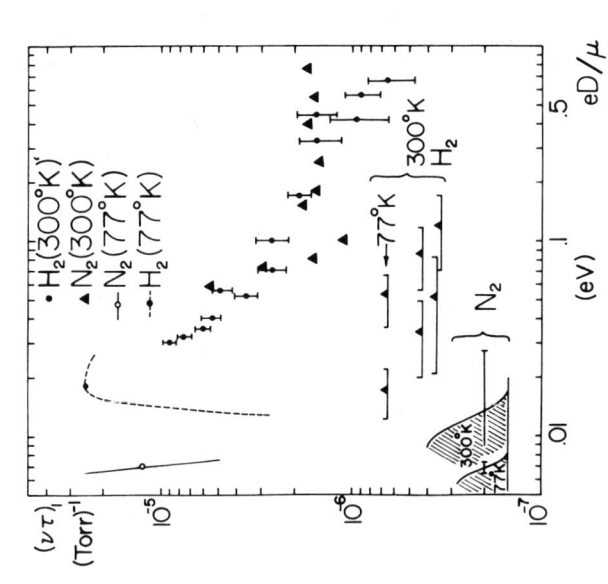

Fig. 2. The slopes $\alpha=(\nu\tau)_l$ of the experimental v_{e0}/v_e versus p curves (as in Fig. 1) for hydrogen and nitrogen at 77 and 300°K, as function of the characteristic electron energy. (The curve for hydrogen at 77°K is somewhat schematic). In a schematic way, the significant rotational resonance levels and their widths are indicated for each gas and temperature.

the resonance collision frequency and is proportional to the pressure independent τ. A non-vanishing slope of the data in Fig. 1 obviously means $\nu\tau \neq 0$. From the constancy of the slope $(\nu/p)\tau$ is obtained for every mean energy of an experiment.

The collision frequencies at unit pressure, ν/p, times lifetime τ, thus obtained are plotted in Fig. 2. We note that at certain energies we have large $(\nu/p)\cdot\tau$, or a strong pressure effect (indicating many resonant collisions at unit pressure, and/or a long trapping time); at other energies the effect is weak or non-existent. There are gases which do not show a pressure effect at all, helium for example.

So far, we have seen that resonance scattering, or the temporary trapping of electrons, can have a measurable effect on the drift motion of electrons. Before I discuss my point, the low-energy resonances, I like to briefly draw your attention to the higher energies, where in hydrogen and nitrogen the well-known single-particle resonances are known to exist (the same resonances Professor Herzenberg discussed in this Conference.[5]) These resonances are known to have a very broad scattering cross-section, peaking at about 2 eV. The width is about 3-4 eV. Hence, at the highest energies of about 1 eV of the experiments here (Fig. 2) a trapping due to the single-particle resonance is to be expected. As can be seen in Fig. 2 to the right, it is also observed. It is found to be of the expected magnitude as well. This can be seen by using the known lifetimes τ and estimating the resonance collision frequency according to

$$\nu/p = N_1 \int v\sigma_r(v) F(\vec{v},\bar{v}) d^3v$$

Herein $\sigma_r(v)$ designates the resonance cross-section, which we obtain in an approximation by subtracting a smooth elastic cross section from the measured total cross section. $F(\vec{v},\bar{v})$ is the velocity distribution function which we assume to be approximately Druyvesteynian or Maxwellian. $N_1 = 3.6 \times 10^{16}$ cm^{-3} is the gas density at 1 torr. In this way one can reproduce these high-energy data points quite nicely. The point I wanted to make in conclusion of these introductory remarks is this: the pressure effect can certainly be understood well in the case of H_2 and N_2 at the highest energies of the experiment.

Now to the real topic of this talk, which concerns the low-energy data: if one uses the above estimate for $(\nu/p).\tau$ to predict what Professor Herzenberg's resonances would do, at the low energies, one finds decreasing ν/p values with decreasing electron energies. As a matter of fact, these known resonances should be insignificant at energies below about 100 mV. What we see from the experimental data in hydrogen and nitrogen, Fig. 2, instead is a sharp increase of $(\nu/p)\tau$ with decreasing energies. It is my opinion that this unexpected increase indicates the existence of as yet unknown resonance states, which appear at energies too low to be accessible in beam experiments. In order to study this low-energy behavior, we look at the most recent data, Fig. 3, by Crompton and Robertson[6]. Plotted is again $(\nu/p)\tau$ versus mean electron energies as in Fig. 2, with more details at the low energies. The data in Fig. 3 are the best data we have at the moment. The gases used in Dr. Crompton's work are n-hydrogen, p-hydrogen, deuterium and helium.

Let's start our discussion with normal hydrogen.

We do not have a pressure effect at all at the lowest energies, but somewhere near 15 mV we start getting one. We find a maximum effect near 30 or 40 mV. If we go to parahydrogen, we find a similar situation except that the whole curve appears to be shifted to the left some. Again, at the lowest energies there is no pressure effect, but we find a steep increase and a peak near 20 mV or so. Interestingly, the isotope ($n-D_2$) shows even at the lowest energies a very strong effect. In helium there is no pressure effect at any energy below 1 eV. These observations can all be understood at once by looking at the threshold for rotational excitation of the various species.

The rotational constant B for hydrogen is 7.5 mV. For normal hydrogen, the predominant rotational transition (at the temperature of $77°K$) corresponds to $J = 1$ to $J = 3$, with a threshold of $10B \cong 75$ meV. According to D. J. Kouri, a rotational resonance occurs at an energy below this threshold. By careful analysis of the shapes, Fig. 3, Crompton and Robertson determine a resonance at 60 mV associated with a virtual $J = 1$ to $J = 3$ transition. (The difference of approximately 15 mV between threshold and the position of the resonance corresponds to the electron affinity of the $J = 3$ state). In parahydrogen

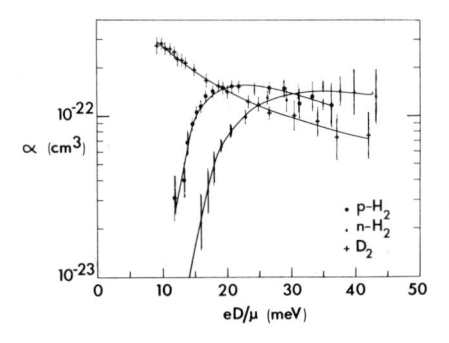

Fig. 3

at 77°K, the predominant transition corresponds to j = 0 to j = 2, with a threshold at 6B = 45 mV. Crompton determines 37 mV as the position of this resonance (thus indicating an electron affinity of about 8 mV associated with J = 2 rotational state). Owing to the inverse mass dependence of the rotational constant B, the isotope's B is the only one half of 7.5 mV, or 3.75 mV. Therefore, in n-D_2 with a prevailing J = 0 to J = 2 excitation, the threshold lies at 22.5 mV. With an electron affinity between 8 and 15 mV we estimate the position of the resonance in D_2 near 10 mV and, on account of the apparent width of the resonances one is not surprised to find a strong pressure effect in deuterium near 10 mV where in hydrogen strictly the absence of a pressure effect was observed. Unnecessary to say that in helium no "rotational resonances" exist owing to the absence of rotational states.

Summarizing it can be said that from swarm experiments, strong evidence is derived for the existence of new, low-energy resonances in electron-molecule scattering. Electron trapping is clearly related to the various thresholds for rotational excitation, and the linear pressure dependence of the observations suggests a two-body compound state. Most recently W. Raith and J. E. Land[7] have attempted a beam-type experiment to look for these resonances. In spite of the tremendous difficulties of such an experiment they were able to present at a recent meeting in Boulder preliminary results in favor of the existence of rotational resonances. We feel that Crompton's and Raith's experiments give strong experimental evidence for the existence of this new class of low-energy resonances, called "rotational resonances."

TEMKIN: Dr. Henry will tell us about the calculations they performed, which I think are on the best theoretical basis that we now have for doing such calculations.

HENRY:

I am going to start from where Dr. Frommhold finished. In addition to swarm experiments there has now been some recent electron transmission experiments performed by Raith and Land[7] at Yale, and this is some of the preliminary data (see Fig. 4) which they presented last summer at a conference in Boulder. What they have given here is for molecular hydrogen and deuterium. The top

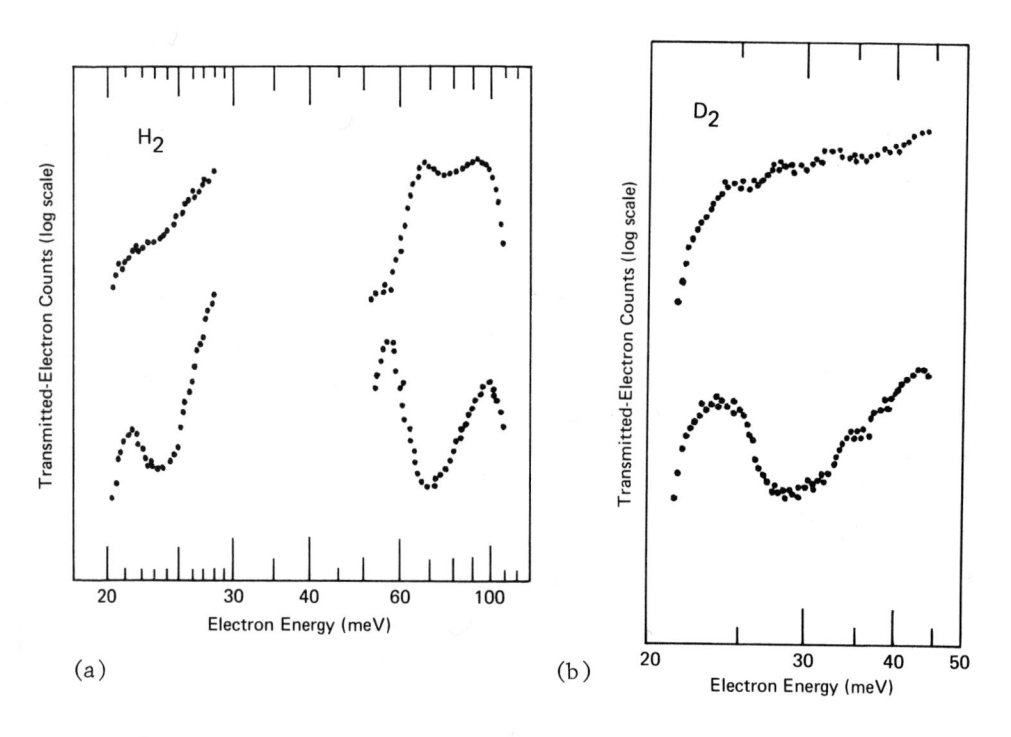

Fig. 4. Electron transmission spectra for H_2 and D_2. Lower curves were obtained with drift-tube pressures of ~10^{-3} Torr; upper curves, with the drift-tube pressure reduced to ~10^{-5} Torr. The horizontal axes are linear in flight time.

part here is taken at a pressure of about 10^{-3} Torr and the lower curves are taken at a pressure of about 10^{-5} Torr. I've tried to pencil in where the various rotational thresholds are (they are at about 44 meV and 72, and so on). Their preliminary analysis is that they appear to be getting a resonance in molecular hydrogen here at 24 meV, and that seems to be fairly well established by their data. They say that the region from about 60 to 90 meV they have a rather broad resonance, if you want to call it that. Originally they had thought that they may be seeing some sort of resonance of 70 meV, but that is purely tentative as of now. The structure that they have in deuterium, in fact, is rather different and their interpretation is that they may have a series of resonances occurring in the region from about 20 to 40 millielectron volts, but the importance is that here we have two entirely independent types of experiments. The first type was a swarm experiment performed at reasonably high pressures. You are dealing with number densities in the order 10^{19} or 10^{20} cm^{-3}. Now we have an electron transmission experiment and the time of flight spectroscopy experiment, and in this the typical densities are around about 10^{14} cm^{-3}. So, it is not just purely a high pressure effect that is going on. There must be something else.

Now, we come to the more negative aspects of the theory, but there is truly very little to say from a theoretical point of view. I first of all would like to just give very briefly what are thought to be very good energy curves based at least, on a fixed nuclei type of approximation. Essentially, these curves come from Taylor and some of his co-workers.[8] They calculated the potential energy versus the inter-nuclear distance for the separation between the two hydrogen atoms. The lowest energy curve that you got for the hydrogen molecular ion is somewhere around 2 eV above the ground state energy of H_2. Now, to interpret the swarm experiments, the electron and the molecule have to get together and form some sort of complex at somewhere around maybe 24 meV. That is basically if you like, one piece of negative theoretical evidence. The next thing I'd like to show is the sort of interpretation one could give to the resonances that occur. For example, if you do a calculation you would first of all expand the wave function for the system, that is the system of the electron plus the molecule, you would expand this in terms of maybe the molecular states of the system in the various rotational levels of J, and if you isolate this partial wave expansion, you can look, for example, at one particular process where you have a d wave electron collides with molecular hydrogen which is in its lowest rotational state, the J = 0, and then you can think of it as if you have enough energy to cause the rotation, then you have an outgoing s wave electron given by $\ell' = 0$, leaving the hydrogen value caused the rotation to occur from the J = 0 to J = 2 state, and we've just conserved the angular momentum going across this. If, for example, we had a d electron coming at some energy that is going to be resonance energy. The d electron gives up its angular momentum to the hydrogen molecule. It now becomes an s electron. It says I would like to go out on that channel with the hydrogen molecule in the J = 2 level, but it says I can't. I don't have enough energy, but it does hang around a little bit. It gets time delayed, and then it says: "whoops" I can't really make it, let me go back the way

I came: that is a rather naive sort of picture, rather simplified thing. So, the essence is: there is a transfer of angular momentum to the molecule, and then it has to give its angular momentum back. We get the d electron back out again, with the hydrogen still in its J = 0 level, and to say that is a reasonably plausible basis on which to try and build a theory. As far as the theory is concerned: Neal Lane and I[9] did some calculations several years ago, and we put in the various effects. We put in the quadrupole moment effect, the polarization effects, electronic exchange effects; we tried to construct a reasonably good potential interaction between the electron and the hydrogen molecule. We used this potential primarily to obtain rotational excitation cross sections which give very good agreement not only for the over-all total cross section, but also the angular distributions. Thus, on this particular potential, we then tried to search for resonances below the J = 2 and J = 3 levels; that is below the 44 meV and below the 72 meV thresholds, and we find that we weren't in fact able to get any resonances. We just couldn't pick up the state. So what it requires is some small tickering around of the potential in order to still give us a potential which will give us the good agreement. By good agreement we mean we were fortunate in getting the agreement within about two percent with the experimental results on the rotational cross sections at the higher energy. So, it is a question of can you play around with the potential - if you still have got very good agreement with the rotational cross sections and at the same time obtain rotational resonances. And we concluded at that time that, in fact, we couldn't do that.
TEMKIN: The essence of the matter right now is that the

possibility of Feshbach resonances associated with the virtual excitation of a J = 2 level in hydrogen or something like that would be a reasonable mechanism - rotational close coupling, at least at the present time, do not seem to show them.

HERZENBERG:

This is an absolute desperate attempt to try and get a long live resonance out of that potential energy curves Dr. Henry just showed them.

From the lowest shape resonance in H_2 you didn't seem to be able to get any sort of resonance at energies less than 2 electron volts. Now, this is potential energy curve picture, of course, which holds if you have no angular momentum. Let's suppose we have some angular momentum in the molecule, in other words let's just add the centrifugal energy, then to both these curves here you would have to add something which goes as $\ell(\ell+1)/R^2$ multiplied by $\hbar^2/2m$. It is quite obvious that if you take each of these curves and add to it the angular momentum what you will get is new curves which have the minimum displaced to the large R. In principle, it is always possible to find an angular momentum rather uncomfortably large, which makes the minimum in this curve come here or even to the right, at this point. So, in that case, in fact, the H_2 curve relative to initial state of H_2 with angular momentum would have been moved down to zero energy. So, if you had a hydrogen molecule with a high angular momentum to start with, you will always find an angular momentum which, in fact, will give you zero resonance - on top of that, of course, because the resonance has not moved down to zero energy.

MASSEY: I would like to make one comment about the possibility of that H_2^- curve being down at the bottom.

I think this is really wrong, because there is experimental evidence on the scattering of electrons by H_2 taking a total cross section on the angular distributions which show that the corresponding phase shift is such that that curve would have to be more or less where it is as calculated by Taylor.

LANE: One point which is not original with me at all but has been raised by several people is that the assumption was made in all the calculations of the polarization potential, which have gone into the calculations by Ron Henry and myself. The assumptions made there were that the recipe used in the polarized orbital treatment of Temkin which appears to be correct in helium, namely, that you shouldn't assume adiabaticity all the way in. You shouldn't assume that the system responds adiabatically to the electron - no matter how close it is to the nucleon, because even if it is a zero energy electron it accelerates. So, we did use that recipe in constructing our polarization potential. Now it has been suggested that since this is a diatomic molecule and not helium, and when one comes to the center of molecule he is not at the nucleus in fact, but between them, that perhaps the nature of this correction, which is essentially the same non-adiabatic correction that Professor Spruch[10] eluded to earlier, is different, and the polarization potential might be adiabatic farther in, and it will get very deep if that is the case. Now I would like to say I don't believe that, because there's no way that Ron and I or anyone else, I would challenge, can get decent angular distributions and elastic and rotational excitation cross sections using such a potential. D. J. Kouri[11] on the other hand made a phenomenological approach using very short range square wells, which are quite deep and there they are deep enough to contain these resonant states that are discussed. On the other hand they are no where near in shape or properties of any other kind the potentials that Ron and I obtain in using reasonable ground state electronic wave functions for H_2. So again, I think that's unrealistic, but perhaps the point that didn't get made here is that, in fact, model calculations have been done with square well potentials which do give reasonable scattering results and do contain these resonant states.

TEMKIN: Well, I would just like to make one comment on that. I think that one of the things that Professor Spruch was saying yesterday was that in a sense you can use the adiabatic idea to bound the phase shifts with the scattering from the other end. In other words: the adiabatic criterion in general will give you a bigger phase shift and so it seems to me, in effect, an artificial increase of the potential, which in some senses is already calculated from the adiabatic approximation, is in the opposite direction from what the bounds seem to indicate. It is very difficult for me to understand how non-adiabatic effects could give you the resonance. That is my own understanding. I'm going to stop the discussion, although it is a very interesting one.

TEMKIN (Adiabatic-Nuclei Theory of e-N_2 Scattering)

I would like to just make a few remarks which may overlap some of the things that Arvid Herzenberg said. The point is that for calculating most aspects of electron-molecule scattering one can artificially define a problem in which the nuclei are fixed and their internuclear axis makes angle $\underset{\sim}{\beta}_0$ with respect to the incoming

ELECTRON MOLECULE COLLISIONS

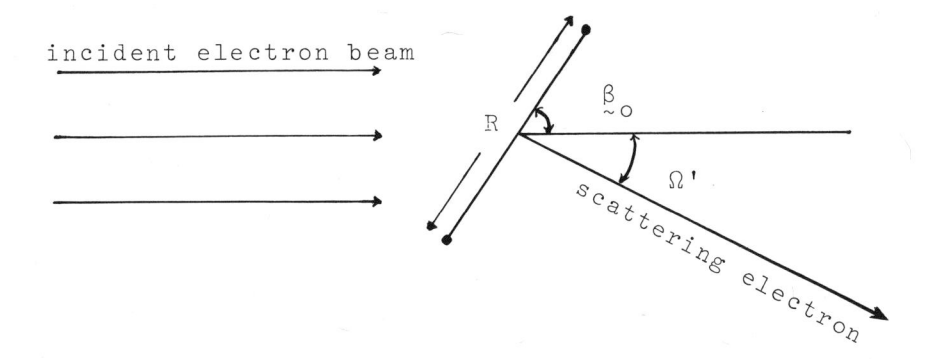

[Figure: incident electron beam scattered at point R by target, with angles β_o and Ω'; scattering electron exits at angle Ω']

beam, and where instantaneously the nuclei are a distance R apart, and then one can construct therefore an amplitude which is a function parametrically of $\underset{\sim}{\beta}_o$ and R given thus: Let

$$f(\underset{\sim}{\beta}_o, R; \Omega') = \text{fixed nuclei scattered amplitude}$$

then

$$f \alpha \sum_{\ell m'} \{\sum_m e^{i\eta_{\ell m}^{(R)}} \sin \eta_{\ell m}^{(R)} D_{m'm}^{(\ell)}(\beta_o) D_{om}^{(\ell)*}(\beta_o)\} Y_{\ell m'}(\Omega')$$

Now, the point is the following: that in the Born-Oppenheimer approximation, that is in the approximation that one considers the nuclei fixed, it turns out that the dependence on the angle $\underset{\sim}{\beta}_o$ can be taken out analytically from the expression in the amplitude, whereas the dependence of the phase shifts $\eta_{\ell m}$ are parametrically functions of R. One can define a fixed nuclei cross section simply by treating $\underset{\sim}{\beta}_o$ as if it were a classical angle, and averaging the scattering from all directions, and what one arrives at is an expression which is extremely like the ordinary scattering expression one has in electron-atom scattering except that now the phase shifts depend on the magnetic quantum numbers, in addition to the orbital angular momentum quantum number corresponding to the fact that the target system is not spherically symmetric

$$\sigma_{\text{fixed nuclei}} = \frac{4\pi}{k^2} \sum_{\ell,m} \sin^2 \eta_{\ell m}(R_o)$$

Consider the transition amplitude between two vibrational and/or rotational states, i.e:

$$f_{\Gamma'\Gamma} = \frac{1}{4\pi} \int e^{-i\mathbf{k}'\cdot\mathbf{r}} \Psi^*_{\Gamma'}(\xi) \, V \, \psi_{k\Gamma}(\mathbf{r},\xi) \, d^3r \, d\xi$$

An exact expression for the scattering amplitude can be written down, and this is the analog of what Dr. Herzenberg showed you before, and this involves the exact wave function $\psi_{k\Gamma}$ from an initial state Γ where Γ represents a state of definite rotational quantum numbers and vibrational quantum numbers of the target and k the incident electron's momentum to a final state $\Psi_{\Gamma'}$ of the nuclear motion. Now the Born-Oppenheimer approximation, which I believe is really behind all this says the following, as I understand it, in this context: that to a very good first approximation you can write the total wave function in a separable form in the sense of a product of something which depends on the position of the scattered particle times wave function describing a nuclear system described by coordinates ξ. If you make this assumption and you put it back into the original integral equation for the scattered amplitude, you get

$$\psi_{k\Gamma} = w_k(\mathbf{r},\xi)\Psi_\Gamma(\xi) + O[(\tfrac{m}{M})^\nu, E, \Delta E]\} \; .$$

The expression in square brackets is nothing but the integral form of the scattered amplitude for the fixed nuclei amplitude. Therefore, if you realize that, then you can write down the fact that the matrix element in the so-called adiabatic nuclei approximation[13] (and I should indicate that this is essentially the same thing that Dr. Herzenberg called the impulse approximation) is a matrix element of the flixed nuclei amplitude function of the scattered angle Ω' between initial and final rotational and/or vibrational states, whose collective quantum numbers are called Γ and Γ', plus an error term. In short:

$$f_{\Gamma'\Gamma} = \int d\xi \Psi_{\Gamma'}(\xi) \left[\frac{1}{4\pi}\int d^3r\, e^{-i\underline{k}'\cdot\underline{r}} V(\underline{r},\xi) w_k(\underline{r},\xi)\right] \Psi_\Gamma(\xi)$$

$$+ O[(\tfrac{m}{M})^\nu, E, \Delta E]$$

But $f(\underline{\beta}_o, R; \Omega') = \frac{1}{4\pi}\int d^3r\, e^{-i\underline{k}\cdot\underline{r}} V(\underline{r},\xi) w_k(\underline{r},\xi)$

$$\xi \equiv (\underline{\beta}_o, R)$$

and $\qquad |\underline{k}'| = |\underline{k}| + O(\tfrac{m}{M})$

Therefore

$$f_{\Gamma'\Gamma} = \int d\xi \Psi_\Gamma(\xi) f(\xi,\Omega') \Psi_\Gamma(\xi) + O[(\tfrac{m}{M})^\nu, E, \Delta E] \quad .$$

Now, in vibrational excitation which is what I want to talk about, let us assume that the experimental resolution is such that it cannot discriminate between rotational levels, but it can discriminate between vibrational levels - so summing over final rotational levels, then it turns out that the total cross section is of the form:

$$\sigma_{v'v} = \sum_j \sigma_{v'j';vj} = \frac{4\pi}{k^2} \sum_{\ell,m} |t_{\ell m}(v'v)|^2$$

where

$$t_{\ell m}(v'v) = \int \chi_{v'}(R) e^{i\eta_{\ell m}(R)} \sin\eta_{\ell m}(R) \chi_v(R) R^2 dR$$

The essential quantity being $t_{\ell m}$: a matrix element between initial and final vibrational wave functions $\chi(R)$ of $e^{i\eta_{\ell m}} \sin\eta_{\ell m}$ where the phase shifts are parametrically functions of R.

At this point the relevant question to ask is what are the conditions of the accuracy of the Born-Oppenheimer approximation itself. One condition on which we all agree is that the electron velocity should certainly be large compared to the nuclear velocity. It turns out for the electron-nitrogen resonance that the energy is at about 2.3 electron volts this is about one half of an atomic unit of velocity whereas the nuclear velocity is about $\frac{1}{2} \times 10^{-2}$ a.u. So that criterion is certainly well satisfied; the other condition that is usually discussed, a condition on which I don't know whether Dr. Herzenberg and I are in complete agreement, is the amount of time that the electron is attached to the nitrogen and whether that is small compared to the transit time. Here it seems to me there are two times characterized by the gross or fine structure width. The gross structure width from the figure gives 10^{-15} sec. which is about the same magnitude as the time of transit time, if you say 10 Bohr radii is the distance of the average potential that electron must traverse. However, if you take the fine structure width then you get even a longer lifetime, 10^{-14} and the ratio is not so favorable. But the fine structures are not completely well developed resonances, so I am not sure how important that is. However, this whole question of lifetimes vis a vis the Born Oppenheimer approximation is a tricky one which I think can early be distorted by this kind of comparison. I hope to be able to discuss it in more detail elsewhere. For now I will just say that I believe that the adiabatic nuclei approximation should at least qualitatively be able to explain the entire experimental structure. In Fig. 5, I show on the right hand side the vibrationally

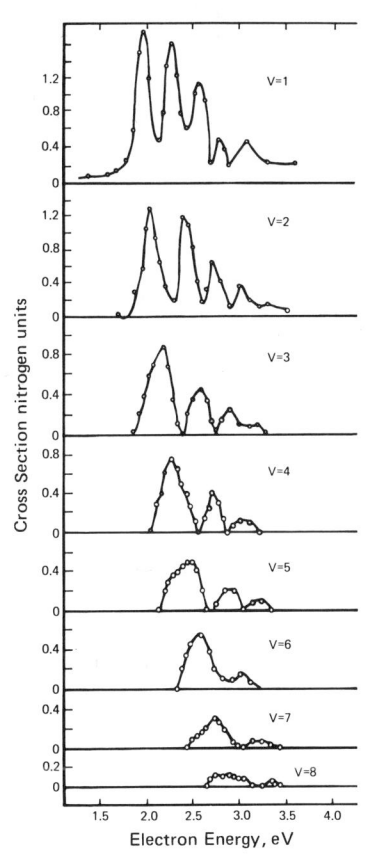

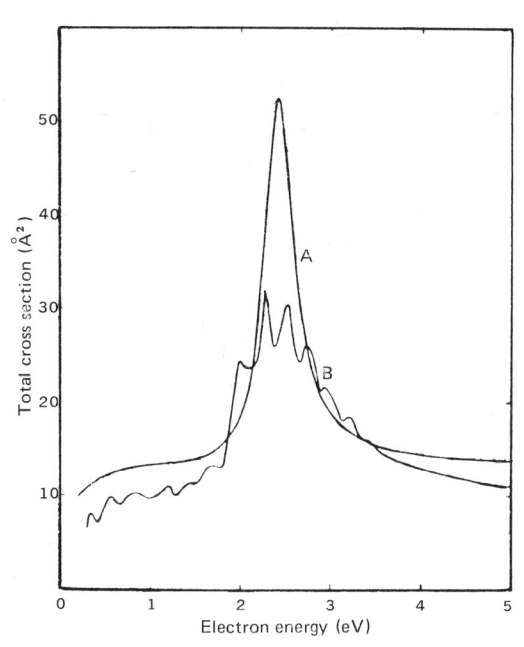

Energy dependence of the vibrational cross section of nitrogen by electron impact. The curves are obtained from sets of curves similar to Fig. 3 taken at different energies of the incident electrons. When the ordinate numbers are multiplied by 10^{-16} a cross-section scale (in cm^2) is obtained such as to give a total vibrational cross section in agreement with Haas. See text for a discussion of errors in cross-section scale. (Ref. 16)

The total cross section; comparison (A) theory with (B) the experimental measurements of Golden (1966). (Ref. 15)

Fig. 5

elastic scattering cross section (B) in an experiment by Golden[14] and these sub-peaks are the fine structure. Curve A is a fixed nuclei calculation:[15] that is calculating the scattering at fixed equilibrium separation. The formula for that cross section in the uncoupled approximation, I have given as $\sigma_{\text{fixed nuclei}}$ before. The figure shows that you can get the gross structure reasonably well from the fixed-nuclei approximation, and so the question is can one get the fine structure from the adiabatic-nuclei approximation?

On the left of figure 5 we show some of the old results of George Schulz[16] for vibrational excitation of N_2 going from the ground to the first, second, third, etc. levels and you see in all cases quite clearly that they also show the fine structure, and therefore, we must naturally demand that that too be explicable from the adiabatic nuclei theory.

Since our work on this is only in progress, I cannot give the final theoretical answer at this time. So let me indicate to you the course of our present thinking and probing. Although the resonance in question has been reliably shown by Krauss and Mies[17] to be a shape (and not a Feshbach) resonance, the fixed nuclei calculation of Burke & Chandra indicates that the resonant Π_g ($\ell=2$, $m=1$) partial wave does increase by π radius: this suggests the usual form given by:

$$\{\eta_{21}(k^2) = \eta_{21}(k^2)_{\text{(non-resonant)}} + \tan^{-1}\left[\frac{\frac{1}{2}\Gamma}{E_r - k^2}\right] \quad (1)$$

where Γ is width and E_r the (fixed nuclei) resonant energy. Both Γ and E_r can depend on R and the first question we asked is does such a dependence on R explain the fine structure? Bertwistle and Herzenberg[18] have inferred that the dependence of Γ and E_r and R is smooth. Therefore we have tried that kind of dependence (and I should emphasize that these calculations are being done in collaboration with L. Wijnberg), and we find that we <u>cannot</u> get the fine structure. Next we can try to make Γ and/or E_r oscillating functions of R. And finally if that doesn't work we will abandon the Breit-Wigner form of η_{21} and try something like:

$$\eta_{21}(R) = \eta_{21}(0) + e^{-\gamma(R-R_o-\Delta)^2}(\eta_{21}(R_o) - \eta_{21}(0)) \quad (2)$$

where

$$\gamma \equiv \gamma(k^2) = \text{oscillating function of } k^2$$

and

$$\eta_{21}(0) = \eta_{21}(R=0) = \text{e-silicon phase shift.}$$

By making γ an oscillating function of k^2 we are sure we can simulate the fine structure as well as the gross structure. The test here will be - can the same η_{21} as a function of R and k^2 explain all the inelastic as well as the elastic cross sections, the differential cross section many other experimental data? And finally we must ultimately demand that an accurate fixed-nuclei calculation as a function of R show whatever oscillations our fit with experiment will force upon us.

MASSEY: I would like to make one or two remarks and that is that the question of what one means by the word resonance in all of these cases, and it applies to N_2 as well as H_2 and that is that I think one has to take account rather seriously of the absolute values of the cross section which have been measured experimentally. It is true that the alkali metals may lead the theorists astray, but measuring cross sections for nitrogen is infinitely easier than measuring cross sections for alkali metal collisions, and it seems clear that in both cases - and certainly clear in the H_2 case that if you are going to get agreement with the observed total and differential cross sections (because you've got observations in both of these) that the so called resonance phase shift does not go through $\pi/2$, because the total cross section would then be too large, and one can estimate the amount of the fraction of other phase shift contributions from the angular distributions, so that you haven't got sharp resonances, so that you don't have to worry about something happening within a fraction of an electron volt, and so in hydrogen it certainly seems clear that you don't have the so called resonance phase shift going through $\pi/2$, and in nitrogen, you might notice in that slide that was shown where there was a comparison between a calculation by Burke and Chandra who worked with him but the actual cross section that they get is twice the magnitude of the observed cross section, and here again, if you put in the kind of admixture of other phase shifts and other contributions as observed from angular distributions, then you find that you cannot allow, as far as I can see, this phase shift as being talked about here to go through $\pi/2$, and so one's concept of so called resonance phase shifts would seem to need modification, and what one is really working with, it seems to me, is that you are working on the conditions where you have got not a resonance in the sense of going through $\pi/2$, but where the phase shift simply goes through a maximum corresponding to a delay time, as it were, long enough, as defined by this rate of change of the phase shift for these other considerations to be important, but I don't think that one ought to believe that one has got the resonance in any of the more conventional senses, and I don't think it is of any point to analyze any of these phase shifts into a resonant part and a background part: it is all background in one sense, or all resonance, and you really have to work in that kind of way.

Having said that, I think if one allows this concept of the maximum time delay then you can apply most of these considerations, but it is something which is different from conventional resonances, particularly the shock resonances which one gets in nuclear physics that is Feshbach resonances, and I think this is an important feature of the situation. But thinking of it from the conventional resonance point of view the point would raise considerable problems. I don't see, in fact, how one is going to impose on a picture of this kind a rapidly oscillating value of gamma. This is certainly going to cause trouble, and in fact, one of the difficulties it seems here in understanding the nitrogen situation is the presence of the shock resonances as given from these oscillations and peaks, and the fact that they are contained in the cross sections for vibration excitation and also, I think, for the elastic resonance part of it, so called, is quite small compared with the total cross section, and so you've got a lot of problems involved here, which I don't think are resolved, and I do believe, for instance, the calculations that Burke and his colleagues have done are very, very interesting, but I don't think they bear too closely on many of these particular issues. They are good old style calculations: old in the sense that I can appreciate them.

Bullard and I[19] were the first to measure the angular distributions in nitrogen, and so I am fairly familiar with them, and also with the analysis of nitrogen data, I can remember doing that in 1930. However, I very much hope that what Dr. Temkin has been suggesting here, that his method of calculating these vibrational and rotational vibrations is valid, because I did, in fact, write a paper published in Phil. Mag. about ten or 12 years ago in which I also proposed that particular method, but I never could apply it at the time because the calculations of phase shifts with nuclear separation were not done at the time, but I do feel that this problem is still one which is important yet presents very great difficulties, and I do feel that this whole question: the relation to the time delay concept to the modes of calculation is important, because what it would really mean is that if for example you take that H_2^- potential energy curve to find that point on the curve with any particular value of R defines the energy at which the phase shift for scattering of electrons by molecule, when a nuclear separation has that value, has a maximum rate of change at that energy, and it is quite a nice and interesting mathematical problem to show how the method of stabilization of eigenvalues of H.S. Taylor does in fact give you this particular feature. I think this is the important part of physics concept of the picture.

I don't think I contributed very much except in a

somewhat negative way to some of the remarks, but I do think it points to some of the difficulties.

TEMKIN: If I could just say one or two words about Professor Massey's remarks. There is nothing that forces us to take the resonance shift to go through $\pi/2$, it simply motivated by the results of Burke and his collaborators.[4] This is clear from Equation (2) which would allow the phase shift simply to be enhanced in some way and come down, and it could be that for smaller R's the enhancement is much less and for larger R's it is much greater to the extent that, in fact, it does at some R begin going through $\pi/2$, and it could be that the change in variation from this kind of shape to a more traditional kind (as far as the phase shift goes for various R's) is precisely what one needs to get this additional variation in the cross section or the fine structure. That is one possibility implicit in the above equation.

Let us just make the assumption that one can do a consistent calculation at each R for the fixed nuclei phase shift. Can one then with an additional quadrature calculate or not calculate the scattering cross section? That is really the essential question, and what we are trying to do now is simply to paramatrise what we find, but it seems to me that if one can then get agreement with experiment one has a well defined method of calculation, and I think that is what Professor Massey and I are primarily concerned with.

HERZENBERG: First of all let me say something about getting the funny detail fine structure in nitrogen out of what Dr. Temkin calls the adiabatic approximation. We've quite agreed about what that approximation is -- the calculation you have to do. You have to work up an integral over a nuclear separation here over a final vibrational state from an initial vibrational state, and in between you have some electron scattering amplitude here which depends on R, and, I supoose, the incident electron momentum. Now, I think one thing is clear: that if $\Gamma(R)$ and $E_r(R)$ were, infact, smoothly varying functions of R then I think it is very, very difficult to see how you would get that sort of oscillation out, and therefore, this is your argument for saying that maybe we ought to look for possible oscillations in these. And that seems unlikely to me.

G. BREIT: I wonder if you couldn't look back at the development in nuclear physics, and there you have very similar things happening. Now, if you want the general theory, you have a very nice general theory that of the Wigner R matrix; it is a wonderful theory. I mean the mathematics is rigorous and, of course there are troubles that are really relativistic, but so far as the present problem goes, I see no trouble. However, it is a wonderful expansion. You have energies occurring, which if you like you can, all resonance energies. You want to apply it to give a nuclear problem then you have to do tricks, you have various energy shifts which aren't exactly their original Dirac shift, but as Thomas showed, the late R.G. Thomas, you can do that systematically, if that is what you want, but may I now ask a question. Isn't resonance just in the nature of things - a useful approximate concept, and cannot run it to death.

TEMKIN: I certainly would agree with you that the resonance is a useful concept, but what we are really, I think, trying to understand here is whether the adiabatic nuclear formalism is adequate to describe this structure that we see in the experimental results or is it not; and so if that comes down to the question of whether the function $\Gamma(R)$ is a smooth or oscillating function of R then this is a question that should be calculationally answerable. Since as yet nobody has ever done the calculation, of course, one doesn't know; simple minded pictures in one dimensional sense, certainly give the impression that gamma should be a monotonic function of R. However, I'm not convinced that a many-body calculation where $\Gamma(R)$ appears naturally out of a many-body calculation necessarily has to be a smooth function of R.

<u>G. BREIT:</u> May I make just one additional remark. I would like to see if Professor Massey looks at the matter of the essential thing of a resonance just very much the way I do. There is one simple consideration. You have another system, let's say an accoustical cavity. You can induce resonances. You can attach pipes, organ pipes or something to it, or an electrical cavity and attach wave guides. Then you can put in a phase changer in your wave guide. So your frequencies won't change. The position is going to change, but the resonating element which represents the thing one might call physics is a system which has a definite vibrator of some kind and is a mode of oscillation that does not directly produce the phase shift but then you look at sharp resonance. You make a full analysis. You look at the complex plane by shifting the path of integration and making a calculation. You find there is a little pole close to the axis. When you are close to the pole, you move the vector in the diagram and you get the rapid change. That represents the essential physics of the situation: a mode with a definite geometry, and then you have the shifts that occur because they escape through the wave guides, can happen in various ways different diagrams and phase changes, and it seems to me that that is a useful way of looking at it; and the

resonance energy then always, automatically, if they are sharp, will come very close to the poles. And that also has the advantage that if you want to borrow anything out of dispersion relation papers you are in a better position.

MASSEY: That is you have to get consistency between the various measurements. This is the one thing that tends to get lost, in all this argument and that is the experimental data. I think that in the case of these particular gases the experimental data is really pretty reliable and I think it is very important that theorists should not become too worried over the alkali metals because the alkali metals were extremely difficult and in fact, the theory was right and the experiment was wrong and it is not something which is a universal feature. One has got to compare the whole variety of data including angular distribution, differential cross-sections both elastic and inelastic vibrational excitation and the total cross-sections for all these processes and make sure that one isn't grossly wrong in the theories as these are concerned. This is an important factor. This is an additional thing that Dr. Temkin is pointing out.

REFERENCES

1. J. J. Lowke, Australian J. Phys. 16, 115 (1963).
2. R. Grünberg, Z. Physik 204, 12 (1967).
3. R. W. Crompton, M. F. Elford, and A. I. McIntosh, Australian J. Phys. 21, 43 (1968).
4. L. Frommhold, Phys. Rev. 172, 118 (1968).
5. A. Herzenberg, in this Proceeding.

6. R. W. Crompton and A. G. Robertson, Australian J. Phys. 24, 543 (1971).
7. W. Raith and J. E. Land, Abstract of ICAP, Boulder 1972; see also J. E. Land and W. Raith, Phys. Rev. Lett. 30, 191 (1973).
8. I. Eliezer, H. S. Taylor, J. K. Williams, J. Chem. Phys. 47, 2165 (1967).
9. R. J. W. Henry and N. F. Lane, Phys. Rev. 183, 221 (1969).
10. L. Spruch, in this Proceeding.
11. D. J. Kouri, J. Chem. Phys. 49, 4481 (1968).
12. A. Temkin & K. V. Vasavada, Phys. Rev. 160, 109 (1967).
13. D. Golden, N. Lane, A. Temkin, E. Gerjuoy, Rev. Mod. Phys. 43, 642 (1971).
14. D. E. Golden, Phys. Rev. Letters 17, 847 (1966).
15. P. G. Burke & N. Chandra, J. Phys. B 5, 1696 (1972).
16. G. J. Schulz, Phys. Rev. 135, A988 (1964).
17. M. Krauss & F. H. Mies, Phys. Rev. A 1, 1592 (1970).
18. D. T. Birtwistle and A. Herzenberg, J. Phys. B 4, 53 (1971).
19. E. C. Bullard & H. S. W. Massey, Proc. Roy Soc. (London) A133, 637 (1931).

X-RAY ASTRONOMY*

Herbert Friedman

E. O. Hulburt Center for Space Research

Naval Research Laboratory

Washington, D. C. 20375

Nearly a decade of preliminary x-ray probing with rockets and balloons culminated with the launching of the Small Astronomical Satellite SAS-1 (Uhuru) for x-ray astronomy late in 1970. That satellite caps an era of discovery of fundamentally new information that describes a "violent universe" of high energy phenomena. It brings us to a new level of space research with the forthcoming SAS-B and -C missions scheduled for gamma-ray and x-ray astronomy in 1973 and 1974, and then hopefully to the High Energy Astronomical Observatory (HEAO) series beginning in 1977. The observations to date have revealed a host of x-ray sources characterized by x-ray powers a thousand or more times as great as the optical or radio luminosities, with a remarkable range and variety of temporal variability. These discrete sources are seen against a background of diffuse x-rays, which may be generated in interstellar or intergalactic space or in

---

*A Fiftieth Anniversary publication of the Naval Research Laboratory

a myriad of unresolved discrete sources at great distances.

The Uhuru satellite is still functioning almost two years after launch and has already built up a catalog of about 125 sources. Two-thirds of these sources lie close to the galactic plane and are presumably members of our galaxy. The remaining third lie at latitudes greater than $\pm$ 20° and may be extragalactic objects. A substantial number of high latitude sources are identifiable with radio galaxies, Seyfert galaxies, and quasars as well as cluster of galaxies. Prominent among the galactic sources are a number of supernova remnants, including the Crab Nebula, the Supernova of Tycho Brahe (1572), Cassiopeia A, and the Veil Nebula in Cygnus. In their gross emission characteristics, these x-ray sources are relatively stable. Nearly all other discrete sources in the galaxy are characterized by very high variability, from random noisy sputtering down to millisecond time scales to the highly precise periodicities of pulsars and eclipsing binaries. Finally, there have been nova-like x-ray sources which flash to brightnesses greater than any other x-ray object in the galaxy and then fade away in a matter of months.

X-ray emission sources fall into two broad categories, thermal and nonthermal. The former include hot plasmas that produce bremsstrahlung, or black body radiation; the latter include synchrotron emission, generated by relativistic particles circulating in magnetic fields, or inverse Compton scattering in which a collision between an energetic particle and a photon of microwave, infrared, or visible light raises the photon energy to the x-ray range. Many of the galactic sources exhibit x-ray luminosities as great at 10^{36} erg/sec, a thousand times the total power radiated by the sun over all wavelengths. There is evidence of star-like sources within our own galaxy and in Andromeda as powerful as 10^{38} to 10^{39} erg/sec. The total x-ray luminosity of a normal galaxy, such as the Milky Way or Andromeda, is in the range 10^{39} to 10^{40} erg/sec. The powerful radio galaxies, Seyferts, and quasars may reach 10^{46} erg/sec, a thousand times the visible luminosity of a giant spiral galaxy.

SUPERNOVAE AND NEUTRON STARS

The violent death of stars is one of the most fascinating problems of astrophysics. In the supernova explosion phenomenon lies the means of processing all the elements heavier than carbon, and the gravitational collapse that triggers the explosion leads to the neutron star, or black hole. Pulsars are identified with rapidly spinning neutron stars which can generate cosmic rays and x-rays, as well as radio waves, and gravitational radiation which may be detectable here on earth.

Astrophysical models describe a variety of ways in which stars may die. Nearly all the matter may be blown off violently or in repeated outbursts or in a steady stellar wind. The star may contract after exhausting most of its hydrogen fuel until it reaches a white dwarf stage, when electron degeneracy pressure balances the gravitational force. White dwarf densities can reach 10^9 gm/cm^3 and the diameters may be as small as those of the inner planets of the solar system. In a more highly evolved star, gravitational collapse may progress to nuclear density, about 10^{14} gm/cm^3 so that the nucleons are effectively in contact. Neutron degeneracy pressure then balances gravity and we have a neutron star about 10 km in radius and with an upper limit of about 1 to 2 solar masses.

If the star is initially very massive (greater than 2 solar masses), it may collapse beyond nuclear density toward a black hole. The pull of gravity is then so great that no radiation can escape. Only its external gravitational field can be sensed and the most likely possibility of detection is by virtue of x-rays generated when charged particles fall into the black hole.

The Crab Nebula is the most intensively studied remnant of a supernova explosion. It contains a pulsar which, undoubtedly, is the collapsed core of the exploding star that was observed in 1054 A.D. Both the nebula and the pulsar have been observed over the entire spectral range from radio to x-rays, and the pulsations have been detected to energies in the gamma-ray range. The nebula resembles a tangled ball of filaments enmeshing an amorphous mass of gas. A magnetic field of about 10^{-4} gauss permeates the nebula. Relativistic electrons spiral in this magnetic field to produce synchrotron radiation over all wavelengths from radio to x-rays. To produce x-rays in the kilovolt range requires electron energies as high as 10^{14} eV. These high energy electrons radiate so rapidly that their energy is dissipated in just a few years. To sustain the x-ray emission requires a continuing supply of relativistic electrons from some central source. The radius of the x-ray nebula is about one light-year, consistent with the lifetime of electrons shot out of a central source.

Shortly after the discovery of radio pulsars in England in 1967, the Crab was found to pulse at 30 cycles per second. A search for optical pulsations quickly succeeded and identified the pulsar with a faint star near the center of the nebula. In rapid succession infrared, soft x-ray, and gamma-ray pulsations were detected, with the power of x-rays dominating the spectrum. The pulsed x-ray flux (1-10 keV) is a thousand times the pulsed radio flux and a hundred times the pulsed optical flux. All of the evidence fits the model of a spinning neutron star to which is attached a magnetic dipole field of about 10^{12} gauss. The kinetic energy of rotation is being dissipated at a rate of about 10^{38} erg/sec, which is confirmed by a small, but precisely measurable, slowdown. This energy of rotation is converted very efficiently into relativistic particles, which are injected into the nebula to produce the entire synchrotron spectrum with a power some 10^4 times the solar luminosity, more than 900 years after the event that gave birth to the pulsar.

To date, the Crab pulsar is the only one of some 60 radio pulsars that has been also found to pulse at x-ray wavelengths. It provides us with a marvelous opportunity to gain observational data from which to develop theories of condensed stellar objects, and the physics of the extreme state of matter that characterizes them. The present specifications for the Crab pulsar are a radius of only 10 km, a core density of 3×10^{14} gm/cm^3, surrounded by a crystalline crust a few kilometers thick. The inner part of the crust has 10^{16} times the strength of steel, and 10^5 times the conductivity of cooper. Below the crust the core consists almost entirely of neutrons with a small percentage of electrons and protons. The neutrons form a superfluid with near zero viscosity and negligible heat capacity, while the small proportion of protons provide a superconductivity capable of supporting electric currents which can sustain the large magnetic field.

SCO X-1

The brightest x-ray source in the galaxy, and the first to be discovered, is Sco X-1 in the constellation of Scorpius. It has been identified with a 12th magnitude faint blue starlike object, with the flickering optical characteristics of an old nova. Its x-ray flux is also spasmodic and it frequently flares to several times normal intensity in a matter of minutes and may remain in the flare-like stage for several tens of minutes. The x-ray spectrum has the exponential form characteristic of thin plasma bremsstrahlung, with a mean temperature of about 5×10^7 K, but sometimes as low as 4×10^7 K or as high as 10^8 K. Although most of the x-ray emission (1-10 keV) fits a simple exponential spectrum, there is an extended tail to high energies which appears to be nonthermal, and at the very soft x-ray end of the scale there is evidence of opacity. In the simplest model of a spherically symmetrical hot plasma, the radius of the source would be about 10^9 cm and the density about $10^{16}/cm^3$, to match the spectral flux data. The radiative cooling time of such a plasma would be a matter of milliseconds. Energy must, therefore, be supplied continuously to maintain the hot plasma.

One model of Sco X-1 suggests that the energy source is a rapidly spinning neutron star, with a period as short as a few milliseconds, or less. This pulsar is enveloped in a cocoon of gas, which is heated by the particles ejected from the central source. Other models invoke a binary pair with accretion of gas onto a compact component which may be a neutron star or white dwarf. There is no direct observation of Sco X-1 being a binary system and the resemblance to an old nova may be only superficial, but the line of sight may simply be inclined to the orbital plane of the binary system. If the x-ray source is the highly evolved binary companion of a star which fills its Roche lobe, mass may be transferred to the compact source in a sloshing gas stream in such a way as to obscure the occultation. The streaming gas may transfer to a ring around the compact source, which could be a white dwarf or a neutron star, and then subsequently fall onto the stellar surface. Irregularities in this accretion process could account for flaring.

It should be apparent from the preceding speculations that the nature of Sco X-1 is still a mystery. At least one other source, Cyg X-2, may be associated with a similar optical counterpart and two sources, GX-17 + 2 and GX-9 + 1, have similar radio counterparts. Radio detection with the three-element interferometer at the National Radio Astronomy Observatory (NRAO), Green Bank, West Virginia, and with similar equipment in Holland, has been successful in pinpointing the locations of these x-ray sources. In the case of Sco X-1, the radio object exhibits three aligned components, one centered on the x-ray source and the other two spaced about two arc minutes apart on either side. Slight displacements in one of the outer components have been interpreted as associated with changing orientation of particle streams from the x-ray source.

X-RAY SOURCES IN BINARY SYSTEMS

Cygnus X-1 was discovered by an NRL rocket survey in 1965, and when the measurement was repeated a year later its flux had decreased by a factor of 4. Evidence for a wide variability in flux has accumulated with subsequent rocket and balloon observations. With the extended data from Uhuru and recent rocket observations, it has become apparent that the variability has the character of random noise with large fluctuations on scales of milliseconds to tens of seconds. Following the discovery of an associated radio source, the position was established to a second of arc and found to coincide with a spectroscopic binary of 5.6 day period. The central object of this system is a 9th-magnitude B 0 super-giant. Typically, such a star would have a mass of about 12 solar masses and the elements of the binary system would require that the unseen companion have a mass in excess of three solar masses. This mass is too great for a white dwarf. It could be a rather massive neutron star, or possibly a black hole.

No evidence has been found for the 5.6 day period in the x-ray emission, but in the course of about a year-and-a-half of Uhuru observations, it has shown an abrupt change in the average level of x-ray intensity, as great as a factor of 4 in March and April 1971.

Two sources, Cen X-3 and Her X-1, have been identified as binaries purely on the basis of x-ray observations. Cen X-3 has a rapid pulsation of 4.8 seconds period, which is 90 percent modulated. This very stable, fast period is superposed on a 2.087 day occultation cycle. From the Doppler velocity, the mass of the central star is calculated to be at least 15 solar masses. The x-ray emitting object must then have a mass as small as 0.1 solar mass, consistent with neutron star models.

The source in Hercules pulses with a 1.24 seconds period and, like Cen X-3, is essentially fully modulated. It exhibits a 1.700 day occultation cycle. Furthermore, there appears to be a still larger cycle of about 36 days, in which the source is bright for 9 or 10 days and then too weak to be observed for the next 26 days. The Doppler shift of the 1.24 seconds pulsation leads to an estimated maximum mass of 0.2 to 3.0 solar masses for the x-ray source, and 1.2 to 3.2 solar masses for the central object. The optical counterparts of Cen X-3 and Her X-1 have not yet been found.

EXTRAGALACTIC SOURCES

Among the extragalactic sources that have been observed thus far are radio galaxies, such as M-87 (Virgo A), Cen A (NGC 5128), and Perseus A (NGC 1275), the Seyfert Galaxy NGC 4151, the quasar 3C 273, and the Coma Cluster. Most of these were detected prior to the launch of Uhuru, but the satellite has greatly refined the observations and extended the list to include some forty sources which are at galactic latitudes greater than $\pm$ 20°. Perhaps, most of these high latitude objects will turn out to be extragalactic.

Galaxies such as M-87, Cen A, and Perseus A are characterized by active nuclei, in which violent explosive events have occurred repetitively. M-87 exhibits a visible jet, stretching about 4000 light-years from the galactic center and dotted by several dense knots. X-ray observations from rockets have indicated large variability in flux and spectrum over periods of months to years. Uhuru data, however, have not shown any remarkable changes over a time span of months. The x-ray power averages about 10^{43} erg/sec, which is about two orders of magnitude greater than the radio power.

Radio astronomy shows a highly compact source in the core of the galaxy only two light-months in diameter. Inverse Compton radiation, generated by the interaction between relativistic particles in the nucleus and the microwave radio photons, could generate x-rays with a high degree of variability. It has also been speculated that the nuclear activity may result from a single supermassive spinning object of a million to a hundred-million solar masses, from swarms of pulsar-like objects in the nucleus, or from magnetoids of the order of 10^4 solar masses that comprise the visible knots in the jet.

Centaurus A (NGC 5128) is a relatively nearby radio galaxy of a class characterized by symmetrically displaced radio-emitting lobes on opposite sides of the optical galaxy. This configuration implies a series of explosive events that hurled the radio-emitting clouds of relativistic electrons out of the galactic nucleus along the direction of the general galactic magnetic field. A compact radio source has been detected at the center of the optical galaxy, and coincident with the radio center is an infrared hot spot. There are two inner radio lobes about 7 minutes apart and two outer lobes about 3 degrees apart. The x-ray emission appears to come from the central region and not from the extended radio lobes. Since relativistic electrons in the lobes should undergo Compton scattering on the 3 K universal background, the absence of detectable x-ray emission from the lobes permits us to set a limit on the background temperature of less than 5.1 K, if the magnetic field strength is assumed to be 4×10^{-6} gauss (the equipartition value to match the relativistic particle energy). If the background temperature is accepted as 3 K, a lower limit of 1×10^{-6} gauss is derived for the magnetic field. The combination of improved observations in the radio, infrared, and x-ray spectral ranges offers the possibility of defining accurately nearly all of the fundamental parameters of such a complex source.

One or two percent of all spiral galaxies appear to be Seyferts. They are characterized by small, extremely bright nuclei which contain 10^9 to 10^{10} stars within a diameter of about a thousand light-years. The optical spectrum of a Seyfert exhibits strong emission lines of great width that indicate high velocity motions in the gas. NGC 1275 (Perseus A) has both the optical properties of a Seyfert and the radio characteristics of a strong, highly variable radio galaxy. Radio interferometry shows a core and halo structure, with a core diameter less than one light-year. Repeated explosions in the core on a time scale of some ten years release clouds of relativistic particles into the halo, where they accumulate over subsequent millions of years. The nuclear region is also orders of magnitude brighter in the infrared than in visible light, and the newly discovered x-ray emission is about comparable in power to the infrared.

Quasars belong to the family of extragalactic radio sources, but are extremely compact in comparison to normal galaxies. They generally exhibit large red shifts, implying that they are located at great distances. If we accept this evidence of cosmological distances, the energy outputs are enormous. A quasar may appear to be a hundred times as luminous as a normal galaxy and yet occupy only a small fraction of the volume of a typical normal galaxy. As surprising as the intense visible luminosity may be, infrared and x-ray emission can be orders of magnitude greater. For 3C 273, the observed infrared luminosity is about 10^{48} erg/sec and the x-ray power about 10^{46} erg/sec. To supply the energy released by quasars strains the estimates of nuclear fusion energy derivable from the available mass of stars.

Some 2 to 5 percent of all galaxies show evidence of violent energy production in their nuclear regions. Seyferts and quasars may be stages in the evolution of active galactic nuclei. Radio 21-cm observations of our galaxy show evidence for the ejection of matter at the rate of about one solar mass per year in a steady nuclear wind. Runaway hydrogen clouds are observed below the galactic plane. A torus of hydrogen 6000 light-years in diameter encloses the nucleus in the galactic plane and appears to be expanding at about 100 km per second. This hydrogen doughnut alone contains about 200,000 solar masses.

Several clusters of galaxies have been observed in x-ray emission. For the largest clusters, the mean x-ray flux expected simply from the integral of galaxies is at least an order of magnitude less than observed. This x-ray emission may be associated with hot, intracluster gas diffusely distributed. The subject of gas in clusters will be discussed in more detail further on.

DIFFUSE X-RAYS

Underlying the flux of x-rays from discrete sources is a diffuse x-ray background. From 1 keV to 1 MeV the spectrum appears to follow a power law, with a bend in slope near 20 keV. The spectrum is relatively flat with a power index of about -1.4 from 1 to 20 keV. At higher energies the spectrum steepens to an index of -2.3. For the softest x-rays (less than 1 keV), the spectrum is again steeper and considerable theoretical interest attaches to the question of whether this soft component may possibly be thermal radiation from intergalactic gas.

The background at high energies (greater than 20 keV) may be attributed to inverse Compton scattering of cosmic ray electrons on the 3 K universal microwave radiation. The microwave photons have energies of the order of 10^{-3} eV. When scattered by electrons of energies 10^6 to 10^8 eV, the photons are raised to x-ray energies of 10 keV to 1 MeV. If the interaction is confined to normal galaxies, the estimated contribution to the x-ray background must be less than one percent of that observed. But if the relativistic electrons generated in radio galaxies leak into intergalactic space, the Compton scattering on intergalactic gas could produce a relatively large background flux. The process can be enhanced by including an evolutionary factor postulating contributions from sources in remote epochs. Such processes become more important at large red shifts because of the evolution of radio sources and of the microwave background radiation, both of which were more dense in earlier epochs.

The background (1-10 keV) may be composed primarily of the integrated contributions of discrete sources or of truly diffuse x-rays generated in intergalactic space, or some mixture of both contributions. The x-ray power of the Milky Way is estimated to lie in the range 2 to 3×10^{39} erg/sec and the observed power of Andromeda is about 2×10^{39} erg/sec. But, if we assume that all normal galaxies emit similarly and integrate over the volume of the universe out to the Hubble radius, the flux obtained is one to two orders of magnitude less than the observed background.

The x-ray flux from the galaxy could be dominated at any time by a young pulsar. NP 0532, the pulsar in the Crab Nebula, radiates about 10^{36} erg/sec from 1 to 10 keV. If we assume that this pulsar is a typical remnant of approximately one supernova event per hundred years per normal galaxy, we can estimate the universal contribution averaged over time. If a new pulsar is left with 10^{52} erg of rotational kinetic energy immediately after the supernova implosion, about 10^{50} erg of x-rays should become available at one percent conversion efficiency in the lifetime of the pulsar. Averaged over time, this contribution would be sufficient to yield a galactic luminosity of 10^{41} erg/sec and the combined flux from all galaxies would suffice to explain the observed diffuse background (1 to 10 keV). The validity of this estimate depends on the accuracy with which the initial rotational energy can be specified. Some theories prefer 10^{53} erg and others as low as 5×10^{49} erg. Thus far, only upper limit fluxes are available from attempts to detect new supernovae in distant galaxies. These limits correspond to energy limits of 10^{50} to 10^{51} erg.

If normal galaxies cannot provide the background flux, we must next estimate the contribution from radio galaxies, Seyferts and quasars. If 3C 273 is a typical quasar, its observed x-ray power of about 10^{46} erg/sec would lead to an excessively high average background. A similar excess of flux would derive if NGC 1275 were a typical Seyfert. However, it is already known that the Seyfert galaxy NGC 1068, which matches the infrared power of NGC 1275, does not exhibit a detectable x-ray flux within an order of magnitude of that observed for NGC 1275. Our sampling of observed sources is still too small to provide a basis for any statistical estimates. Perhaps those active galaxies which have been detected thus far in the x-ray spectrum are only the most unusual specimens of their classes.

As was mentioned earlier, the x-ray fluxes observed from a few prominent clusters of galaxies such as Virgo, Coma, and Perseus are between one and two orders of magnitude greater than would be expected from the integral of normal galaxies within the clusters. It is not clear how much of the observed flux should be attributed to one or two powerful radio galaxies in each cluster, or to a hot intercluster gas, or some other source. At this time we do not have enough observational information about clusters to gauge their contribution to the diffuse background.

THE VERY SOFT X-RAY BACKGROUND

Observation of very soft x-rays (less than 1 keV) is experimentally difficult and the results must be interpreted cautiously. Nevertheless, the body of evidence based on numerous rocket flights indicates a significant excess of soft x-rays over any simple extrapolation of the higher energy spectrum. Surveys of the sky from the galactic plane to the galactic poles have been interpreted in terms of a two-component model of x-ray sources. First, there appears to be an isotropic extragalactic flux which maximizes at the poles and suffers increasing attenuation by interstellar hydrogen as the angle of view turns toward the galactic plane. Second, there is a component which appears to be due to soft x-ray sources within the galaxy and distributed in the same way as galactic hydrogen. A few such discrete sources have been resolved, for example, the Veil Nebula in Cygnus. This nebula is the expanding front of the debris resulting from an old supernova. At the shock front, created by collision of the nebular debris with the interstellar gas, the temperature rises to a few million degrees and the gas radiates soft x-ray bremsstrahlung.

If we accept the evidence for extragalactic soft x-rays, one possible interpretation is that it originates in bremsstrahlung of intergalactic gas at a temperature of about a million degrees. The required density would be about 10^{-5} protons/cm^3 and would be approximately the amount required to "close" the universe. If, however, the gas is concentrated in clusters of galaxies, the x-ray observations could be satisfied by an order of magnitude less total mass of gas averaged over the universe. This lesser requirement arises because the x-ray emission per unit volume varies with the square of the proton density. It, therefore, becomes very important to establish the amount of hot gas present in the relatively nearby observable clusters of galaxies.

Clusters of galaxies must exert a strong gravitational pull on the surrounding intergalactic gas which should then fall steadily into the intra-cluster regime if the gas temperature is only of the order of a million degrees. Estimates of the lifetime of clusters and the rate of accretion of intergalactic gas could indicate the total density arrived at. Thus, the expected x-ray emission from the cluster would relate to the density of the intergalactic gas. X-rays have been detected from the Coma cluster. If interpreted as bremsstrahlung from hot gas, the amount present in the cluster implies a very low content of surrounding intergalactic gas, far less than the critical mass to provide closure.

CLUSTERS OF GALAXIES

Thus far we have considered the question of intergalactic gas as it relates to the missing mass of a "closed universe," but clusters themselves pose a question of gravitational binding. As in the universe at large, the mass evident in the galaxies of a cluster usually represents only a few percent of the mass required for self-gravitational binding of the cluster.

Galaxies may cluster by the hundreds to thousands. The richest clusters may contain as many as 10,000 galaxies. Finally, clusters may agglomerate into superclusters of as many as a hundred clusters. Near the center of a cluster, the density of galaxies may reach a thousand to a million times the average density of the background sky. Some clusters, such as Coma, are spherically compact and appear to be highly centrally condensed and strongly bound. Others, such as Perseus, are highly irregular, with the bright galaxies strung in a chain. The velocity dispersion of Perseus is very large and the cluster may have positive energy which would make it unstable.

If Coma is bound, the Virial Theorem requires that the potential energy equal twice the kinetic energy of all the components. The gravitational mass, which is derived from the Virial Theorem, is many times greater than is computed from the expected mass-to-luminosity ratio. It has been estimated that the observable mass is as low as 1/7 to 1/3 the binding mass. It is possible that the mass is present in invisible gas, or black holes, or that the cluster is unstable and expanding.

If a binding mass of gas is present, the upper limit on temperature is about 10^8 K, which would give the gas escape velocity. A lower limit of about 3×10^4 K is given by the failure to detect the 21-cm radiation of neutral hydrogen. X-rays have been observed from Coma in the 2-10 keV range, which may be attributable to thermal emission at a temperature of about 70×10^6 K, but the mass of the gas involved would be only about 2 percent of the binding mass. The next question is: "Can the missing mass be present at lower temperatures?" An upper limit on soft x-rays obtained from a rocket observation rules out the possibility of a binding mass of gas being present at temperatures greater than 5×10^5 K. But it is conceivable that gas falling into the cluster from an intergalactic medium at about 10^6 K could collapse into clouds in the temperature range 10^4 to 10^6 K. Collisions of these clouds could raise a small fraction of the gas to 70×10^6 K. If such a cloud structure exists, it should radiate Lyman-alpha (1216 A), following recombination of protons and electrons. An attempt to observe such Lyman-alpha has set a limit which requires that the clouds occupy no more than 6 percent of the volume of the cluster within a narrow temperature range centered at about 2×10^5 K, if the binding mass is present.

MATTER AND ANTIMATTER

If the universe were symmetric with respect to matter and antimatter, the consequences would be readily observable in high energy astronomy, even though conventional optical spectroscopy would find an antimatter galaxy indistinguishable from one built of normal matter. Balloon experiments have failed to detect antimatter in the cosmic rays and the fractional upper limit may be set as low as a few hundredths of a percent. If all the locally measured cosmic rays were confined to the galaxy, we could conclude that there is almost no antimatter in the galaxy. If the cosmic rays are extragalactic in origin, it would follow that the universe at large is unsymmetric. Some small fraction of the observed cosmic rays may be extragalactic in origin, perhaps 10^{-2} to 10^{-3}. It is, therefore, very important to improve the sensitivity of cosmic ray observations and to push the energy range as high as possible, preferrably above 10^{15} eV, where the probability of extragalactic origin is much greater. The search for heavy antinucleons (the iron range) is especially important. Since production of anti-iron as a cosmic ray secondary is extremely unlikely, its observation would provide strong evidence of the existence of antimatter in parts of the universe.

In nucleon-antinucleon annihilation, as many as 6 pions may be produced with roughly equal distribution between positively, negatively and neutrally charged species. The π^os (neutral π mesons) decay into two gamma-rays, with a spectrum that peaks at about 70 MeV and spreads from several tens of MeV to hundreds of MeV. There is weak evidence from the OSO-3 gamma-ray observations for the existence of an isotropic gamma-ray background. Isotropy is interpreted as evidence that the gamma-rays originate uniformly throughout the intergalactic medium or in many discrete sources at great distances. These observations set an upper limit on the annihilation rate, assuming all the gamma rays to be produced in that process. Since the upper limit of neutral intergalactic gas can be set at less than $10^{-11}/cm^3$, the expected gamma-rays from such a small fraction of the material universe, even if it were symmetric, would be far below present detectability.

If the intergalactic gas is fully ionized and comparable to the matter in galaxies, the universal average density would be about $10^{-7}/cm^3$. But if the intergalactic gas were symmetric and fully mixed, its density would need to be $< 3 \times 10^{-10}/cm^3$ (about 10^{-4} of the critical density for a closed universe) or the gamma-ray background would have been detected. If the gas is composed almost exclusively of nucleons, the fractional concentration of anti-nucleons would need to be less than 10^{-5}, if the intergalactic gas density is $\sim 10^{-7}/cm^3$. We must conclude that the intergalactic gas contains a negligible fraction of antimatter or that there is essentially no intergalactic gas at all ($< 10^{-12}/cm^3$).

Satellite and balloon observations have measured a flux of gamma-rays from the galactic plane. The pattern of emission resembles a line source stretching along the galactic plane with an enhanced emission in the direction of the galactic center. The evidence is still somewhat controversial, since some recent balloon observations find point sources near the galactic center but no line source. Just about all the emission observed away from the galactic center can be explained by the process of cosmic ray collisions with interstellar gas, producing π^o mesons which decay into gamma-rays. The galactic gamma-ray emission can therefore be explained without any requirement of antimatter. If the gas were thoroughly mixed, the anti-nucleon concentration could not exceed

10^{-15} of the normal component.

To explain the enormous energy radiated by radio galaxies, quasars, and Seyfert galaxies, the possibility of matter-antimatter annihilation has often been considered. For radio sources, the synchrotron emission could derive from electron-positron pairs gyrating in magnetic fields. Roughly, 17 percent of the annihilation energy would be converted to such particles, accompanied by 10^4 gamma-ray quanta from π^o decay for each erg of particle energy. Accordingly, we would expect to observe intense gamma fluxes from radio sources, but no such gamma-ray point source has yet been positively detected. In particular, the upper limit on annihilation energy set by the observed gamma ray upper limit for NGC 1275 (Perseus A) is about 5×10^{44} erg/sec, yet its x-ray power is about 4×10^{45} erg/sec and its infrared power about 3×10^{46} erg/sec. Similarly, for the quasar 3C 273 we have an upper limit of $\sim 10^{47}$ erg/sec from annihilation and an observed 6×10^{48} erg/sec for the infrared. The absence of intense extragalactic gamma-ray sources, therefore, indicates that strong radio, x-ray and IR galaxies do not derive their energies from matter-antimatter annihilation. If the strong infrared emission from the center of our own galaxy were attributable to annihilation, the accompanying gamma-ray flux would be a thousand times greater than observed.

HELIUM IN SMALL ASSEMBLIES AND HEAVY NUCLEI

M. Rasetti*

Center for Theoretical Studies

University of Miami, Coral Gables, Florida

and

T. Regge

Institute for Advanced Study

Princeton, New Jersey (presented by T. Regge)

In this lecture attention will be focused on the physical properties of small droplets of liquid He^4; "small" refers here to assemblies of a few hundred atoms. The aims and reasons of such an interest are twofold. First, the physical behavior both of a single particle and of a many particle system in its thermodynamic limit--i.e. when it is indefinitely large in size--have been thoroughly discussed in the framework of quantum and statistical mechanics. It remains, however, of wide interest to examine how a system tends to the thermodynamic limit as its size is gradually increased; there

---

*Partially supported by U. S. Air Force Office of Scientific Research under grant No. AFOSR-72-2298.

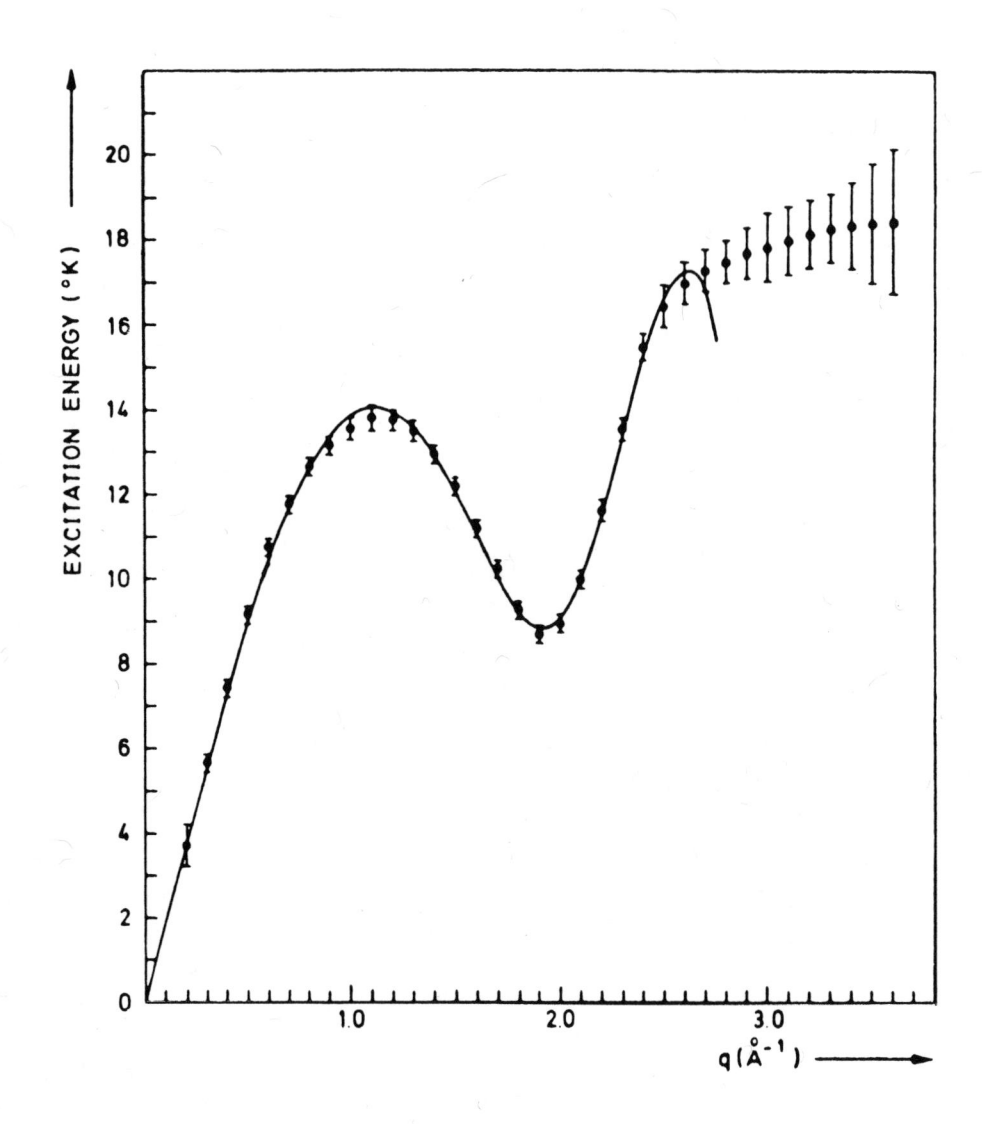

Fig. 1. The dispersion curve in liquid HeII from the experimental neutron scattering data by Woods and Cowley[4].

are also certain deep and peculiar phenomena associated with the small size in the way one sees the trend toward the thermodynamic limit. Secondly, these phenomena—although of a purely speculative nature for the moment—afford some striking and suggestive analogies with heavy nuclei, namely with the liquid drop model of nuclei[1], essentially in the recent formulation by Strutinsky[2] concerning the doubly humped fission barrier.

In liquid He sound is propagated[3], and such a sound is quantized, the quanta being called phonons: the excitation energy spectrum of the phonons* is shown in Fig. 1.[4] The minimum of the curve is referred to as the roton part of the excitation energy spectrum, and it occurs approximately when the wavelength of the corresponding elementary excitation is about equal to the average interparticle separation in the fluid. It is possible to fit the roton part of the curve in Fig. 1 in several ways; Landau[6] proposed his own fit as a parabola. A somewhat different way is to think of it as the upper branch of an hyperbola, namely the square root of a polynomial[7]. In such a case it is possible to extrapolate the function onto the complex-momentum domain, and one easily finds that it has two complex zeros with a small imaginary part.

In other words there exist two complex wavevectors for which the energy of propagation is vanishing.

The motion of liquid He can be described in terms of the velocity field $\vec{v}(\vec{r},t)$ or the velocity potential, $\phi(\vec{r},t)$; and it is governed at zero temperature by the

---

*Actually the spectrum measured by Woods and Cowley is that of the density-density correlation function: the two spectra have been however shown identical by Gavoret and Nozières[5].

Bernoulli equation

$$m\frac{\partial \phi}{\partial t} + \frac{1}{2}mv^2 + V(\vec{r},t) + u(\rho) + \frac{p}{\rho} - \frac{1}{N}\int_\Omega p\, dv = 0 \quad (1)$$

and the continuity equation

$$\frac{\partial \rho}{\partial t} + \operatorname{div} \rho \vec{v} = 0, \quad (2)$$

where $u[\rho(\vec{r})]$ is the average contribution to the total binding energy of a single particle state, $\rho(\vec{r})$ is the local density, and the system of N atoms, each of mass m, is assumed to be contained in a cubic volume Ω, whose edges have length $L = \Omega^{1/3}$; p is the pressure and $V(\vec{r},t)$ some external, time-dependent potential. With the above notation the ground state energy is given by

$$E = Nu(\rho_0), \quad \rho_0 = \frac{N}{\Omega}. \quad (3)$$

Equations (1) and (2) can be linearized by assuming

$$\rho = \rho_0(1 + f), \quad f \ll 1, \quad (4)$$

and it is easily found that f represents a sound wave, i.e. a solution of the wave equation

$$\Box f = \Delta f - \frac{1}{c^2}\frac{\partial^2 f}{\partial t^2} = 0 \quad (5)$$

propagating with a velocity c given by the known Huilier formula[8],

$$c^2 = \frac{1}{m}\frac{dp}{d\rho_0} \quad (6)$$

An unsatisfactory feature of Eq. (6) is that it contains

no mechanism for dispersion, i.e. all phonons propagate with the same frequency-independent speed (actually the given derivation essentially amounts to considering phonons with a very long wave length as compared to the average interparticle distance). This can be amended by taking into account dispersion[9] through the substitution of Eq. (1) with

$$m\frac{\partial \phi}{\partial t} + H([\rho],\vec{r}) = 0 \qquad (7)$$

where the enthalpy H is a nonlocal function of the density

$$\rho(\vec{r},t) = \rho_0 + \sum_{\vec{k}} f_{\vec{k}} e^{i\vec{k}\cdot\vec{r}}, \qquad (8)$$

and, in order to use the correct propagation velocity, one writes

$$H = u(\rho_0) + \frac{m}{\rho_0} \sum_{\vec{k}} [\frac{\epsilon(q)}{q}]^2 f_{\vec{k}} e^{i\vec{k}\cdot\vec{r}}, \quad \vec{q} = \hbar\vec{k}, \qquad (9)$$

where $\epsilon(q)$ is essentially the curve shown in Fig. 1. It is worth observing that the kernel appearing in the configuration-space form (9) has a range comparable with the average interparticle distance, so that its non-locality is actually of no concern. One has now a set of phenomenological equations which should be able to describe the high-frequency propagation inside the fluid.

Choosing the Fourier component $f_{\vec{k}}$ [see Eq. (8)] in the form,

$$f_{\vec{k}} = -\lambda [\frac{q_x}{\epsilon(q_x)}]^2 \delta(q_y)\delta(q_z), \qquad (10)$$

the problem reduces to a one dimensional geometry in which waves propagate along the x-axis. It can be noticed from Eq. (10) that, as expected, such waves are not monochromatic, q_x being smeared out around the roton wavelength depending on the shape of the excitation energy spectrum. Where $\varepsilon(q)$ (henceforth the index x will not be explicitly written unless confusion may arise) has a minimum, the term in square brackets on the right-hand side of Eq. (10) has a local pronounced maximum, and $f_{\vec{k}}$ is therefore peaked at the roton wavelength, where it assumes a Lorentzian shape very closely reminiscent of a Breit-Wigner resonance curve. Phenomenologically one finds, in the neighborhood of the roton momentum,

$$\left[\frac{q}{\varepsilon(q)}\right]^2 \simeq \frac{b_o^2}{c_o^2} \left[\frac{1}{(k+b_1)^2 + b_o^2} + \frac{1}{(k-b_1)^2 + b_o^2}\right] \quad (11)$$

where

$$c_o = \frac{\Delta_r}{\hbar b_1}, \quad b_1 = 1.95 \text{Å}^{-1}, \quad b_o = 0.3 \text{Å}^{-1} \quad (12)$$

Δ_r being the energy gap at the roton minimum. The main consequence of such a behavior is that the inverse Fourier transform of $f_{\vec{k}}$ --essentially $[\rho(x) - \rho_o]$--results in a damped wave, whose damping is proportional to the peak width; the damping itself is about a factor of 3 per wavelength, the wavelength being about 3Å. So the wave is very quickly damped, and after about 10 oscillations it practically disappears.

In order to generate such a wave, one doesn't need to add any energy: it is merely static in nature, and therefore when superimposed on the fluid it remains stationary there. There is a price to be paid, however,

represented by a delta function in the enthalpy H which
appears when (10) and (11) are substituted into (9) to
obtain

$$H = u(\rho_o) - \frac{L\lambda m}{\rho(o)} \delta(x) + \frac{\lambda m}{\rho(o)} . \qquad (13)$$

The simple physical picture connected with this is that
if one follows the wave not in the direction where it
decays but in the direction where it grows, it quickly
grows so much that eventually its amplitude becomes
comparable with the background density of the fluid,
and something must happen there in order to keep the wave
from blowing up. This is the origin of the $\delta(x)$ term
in Eq. (13), whose presence can be realized in two
alternative ways: either at $x = 0$ one applies some
singular external potential or the fluid simply just ends
there, i.e. there is the free surface of the liquid
itself. It is to be noted that the present discussion
is not concerned with mathematical rigor in any of its
features, but nevertheless its conclusions are qualitatively right, and they must be handled and accepted
in this spirit. The solution of the wave equation derived from Eqs. (2), (7) and (9) is:

$$\rho = \rho_o + \frac{\lambda}{c_o^2} - \frac{L\lambda b_o}{c_o^2} \cos(b_1 x) e^{-b_o x} , \qquad x > s > 0 \qquad (14)$$

for the part which evolves inside the fluid, and

$$\rho = A e^{2ax} , \qquad a = 1.14 \text{Å}^{-1} , \qquad x < s \qquad (15)$$

Eq. (14) actually furnishes an oscillating density
damped to the usual bulk value as one moves away from
the surface, while Eq. (15) essentially says that the

fluid is a gigantic bound state of He atoms and there is an exponential decay behavior--which is nothing but the tail of such a bound state--as one moves away from the free liquid surface. Joining the two curves (14) and (15) exactly would imply knowing the equation of state in between, but, relying on the assumed smoothness of $\rho(x)$, one may simply demand the existence of a plane surface $x = s$, such that on it the two functions join so as to make ρ continuous up to the second derivative. This provides a certain shape for the fluid which definitely gives a lower energy than the straight edge, and besides allows a number of interesting considerations. The actual selection of the free parameters A, λ, s which fulfill the above requirement is easily performed, and furnishes a small positive value $s = 0.253$Å (thus providing a pretty good definition of the fluid surface) and

$$\lambda = \frac{z \rho_o c_o^2}{L b_o} \quad , \quad z = 0.77 \tag{16}$$

The resulting qualitative plot in the neighborhood of the free surface of the liquid density or, more exactly, of the order parameter which is the vacuum expectation value of the He field, is given by Fig. 2.[9]

The surface tension can then be computed as

$$\tau = \frac{z}{1-z} \frac{\rho_o m}{b_o c_o^2} = 0.56 \frac{\text{dyne}}{\text{cm}} \quad , \tag{17}$$

to be compared with the experimental value $\tau = 0.35 \frac{\text{dyne}}{\text{cm}}$. One may notice how the theoretical result is off less than 40%, which can be regarded as very satisfactory in view of the crude approximations adopted. Of course surface tension calculations have been presented before

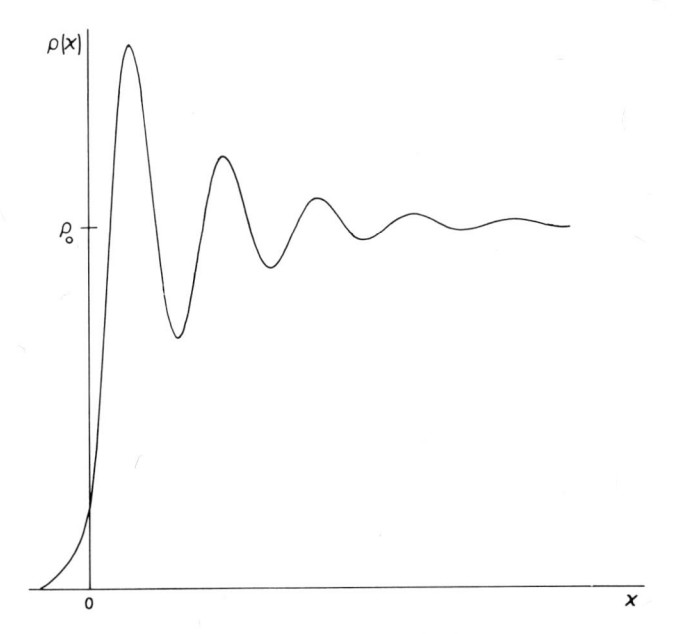

Fig. 2. A qualitative plot of $\rho(x)$ in proximity of the liquid free surface.

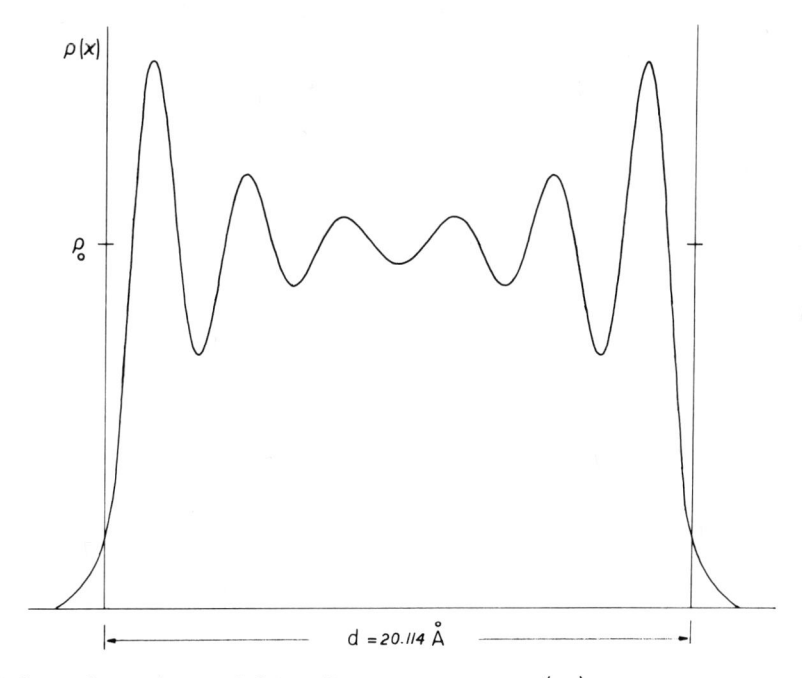

Fig. 3. A qualitative plot of $\rho(x)$ for a slab.

$d = 20.114$ Å

in the literature[10], based upon more sophisticated and accurate models; but the present simple-minded model allows going far beyond the computation of τ.

First one has the picture of this He sea in which the density of the fluid, flat inside the bulk, rises toward the boundary and then decreases to zero with some smoothing on the edges, with oscillations which are sort of bound states of the roton at the walls--a picture very similar to that of Tamm electrons in solid state theory.

Then one may proceed and examine, in the same way, the case in which the He is confined to a slab configuration. If one calculates the total energy in such a case, one essentially obtains three kinds of contributions: the binding energy of the ground state, the surface terms corresponding to the surface tension (17), and an unexpected extra term of mutual energy between the two surfaces of the slab. The latter has the form

$$E_{slab} = 2C \cdot e^{k(s-d)}(k + \frac{1}{d}) + conj., \qquad (18)$$

where d is the thickness of the slab, E_{slab} is the mutual energy per unit surface, $k = b_1 + ib_o$, and

$$C = \frac{z}{1-z} \frac{\tau}{2a} \tau(1 - 2as) . \qquad (19)$$

The operation of summing the complex conjugate in Eq. (18) is due to k being complex, in order to obtain a real energy. Also, because k is complex, E_{slab} is oscillating, and presents maxima and minima depending on the thickness of the film. Such a property can be visualized in the following way: if the slab is thin enough so that the mutual influence of the opposite surfaces is not negligible, the slab itself has the tendency to space its walls in such a way as to be in a minimum energy configuration; which roughly corresponds to the fact that the atoms tend to congregate in layers, and the slab prefers an integer number of layers in between the two limiting surfaces. The maxima and minima of density (see Fig. 3) in fact manifest the high correlation existing among atoms on the walls, this correlation decreasing as one moves toward the bulk. The described effect clearly depends on the finite geometry of the system, and it should be to some extent observable. The immediate further speculation is now: the slab is still infinite in two directions; how do preceding concepts generalize for a finite droplet of any closed shape? One interesting criterion for generalization is to assume that the droplet energy is made up of several pieces: first the sum of the binding energies u of all the particles, then the surface tension term τS, S being the surface of the drop--which are found also in the former liquid-drop nuclear theory[1] (of course nuclei have Coulomb energy which is absent in He)-- and the additional following term:

$$E_u = C \frac{e^{ks}}{2\pi} \int_S \int_S dS_1 dS_2 \frac{d^2}{dn_1 dn_2} U(r_{12}) + \text{conj.} , \quad (20)$$

which is chosen so as to match the specific energy (18) when the closed surface degenerates into the slab. In order to "tune" the layers distribution $U(r_{12})$ must be a solution of the Poisson equation with wave vector k, and its asymptotic behavior is therefore that of a Yukawa potential

$$U(r_{12}) \sim \frac{e^{-kr_{12}}}{r_{12}} ; \quad (21)$$

but k being complex results in a much less effective cut-off action than the oscillating effect.

The double bilocal normal derivative with respect to the surface in Eq. (20) actually makes E_u a sum of pair terms: it is a sort of surface tension depending on pairs of surface elements. The integral on the right hand side of Eq. (20) can be computed exactly for the slab--in which case, as already mentioned, it furnishes $E_u = S \cdot E_{slab}$ [see Eq. (18)]--and as well for a sphere of radius R. In the latter case one obtains

$$E_{sph} = 4\pi \frac{C}{k} e^{ks} \left\{ R^2 U(o) + J - \frac{e^{-2kR}}{k}(Rk+1)^2 \right\} + \text{conj.} \quad (22)$$

where

$$J = \int_0^\infty dx \, x \, U(x) . \quad (23)$$

The first term in curly brackets in Eq. (22) merely renormalizes the surface tension; this means that the contribution to the integral of surface elements very close to each other is a change in the surface tension, giving only a new "effective" τ:

$$\tau_{eff} = \tau + [Ce^{ks} U(o) + \text{conj.}] . \quad (24)$$

The remaining term contains the interesting piece of information because it depends on the shape of the droplet through the relative position of distance surface elements. J does not depend on R, and it turns out to be negligibly small with respect to the last term. The extra energy is therefore widely oscillating, because of the exp(-2kR) factor; and one thus finds oscillations superimposed on the generally smooth trend of the binding energy for the liquid droplet. A similar feature is found in nuclei, i.e. certain nuclei have very high binding energies when they have magic numbers of nucleons. In general nuclei are much more rigid when their binding energies are high, and this would correspond to a minimum in present case.

For the general case one can parametrize formula (20) in a different way, in order to emphasize the meaning of $U(o)$ and the dependence on the shape of the extra term

$$E_u = Ce^{ks} \{SU(o) + J \int_S \frac{ds}{R_{eff}^2}\} + conj. \, , \qquad (25)$$

where

$$\frac{1}{R_{eff}^2} = \frac{3}{8}(\frac{1}{R_1^2} + \frac{1}{R_2^2}) + \frac{1}{4R_1 R_2} \, , \qquad (26)$$

R_1 and R_2 being the principal radii of curvature of the surface. It is now possible to verify that the spherical shape is not necessarily the most stable one, because the term (20) may force a tendency away from sphericity; and this is particularly suggestive because nuclei also do not always prefer a spherical shape for their ground state.

The intuitive description for this is that essentially a droplet, due to this geometry dependent extra energy, tries to assume a shape minimizing its energy. This happens whenever one is able to insert in the droplet an integer number of equally spaced layers (the picture is reminiscent of the shell model for nuclei, even if the shells in nuclei have a different nature, connected to quantum numbers of hydrogen-like wavefunctions). Here one has real space shells which can be

imagined to be built up by putting a single atom in the
center, and around it, one on the top of the other,
successive layers of atoms. One eventually gets a
spherical shape only if the number of available atoms
exactly fills up an integer number of layers: the radius
is in some way quantized, the allowed values corresponding to the minima of E_u. If however, the number of
atoms is not sufficient to complete such a configuration,
the system will have a different shape, acquiring
higher multipole moments because, with a fixed number of
atoms, the above described onion-like structure of layers
is more easily, i.e. with less energy, compatible with
an oblate or prolate than with a spherical shape.

Mathematically this can be shown by expanding the
shape of a slightly deformed droplet surface around the
closest spherical one, in the following way:

$$r(\theta,\phi) = R + u(\theta,\phi) = R + \sum_{\ell>1} u_\ell P_\ell(\cos\theta) \quad , \quad u<<R. \quad (27)$$

Performing now the integration in Eq. (24) over the
surface S defined by (27) one finds

$$E_u = E_{sph} + 4\pi \sum_{\ell>1} (2\ell + 1) u_\ell^2 E_\ell \quad , \quad (28)$$

where

$$E_\ell = [2CR^4 e^{kS}(h_\ell - h_1) + conj.] + \frac{1}{2}(\ell - 1)(\ell + 2)\tau, (29)$$

$$h_\ell = -\frac{1}{2R^2} \int_0^{2R} [xU(x)]^{IV} P_\ell(1 - \frac{x^2}{2R^2}) \, dx \quad . \quad (30)$$

It is worth noticing how the last factor in Eq. (29),
giving just the surface tension, vanishes when $\ell = 1$.
h_ℓ has been explicitly calculated for any $\ell > 2$,[11] but it

is here sufficient to examine the quadrupole term $h_2 - h_1$, given by

$$h_1 = -\frac{k}{R^2}\left(k + \frac{1}{R}\right)\left[\left(k + \frac{1}{R}\right)e^{-2kR} + \left(k - \frac{1}{R}\right)\right] \qquad (31)$$

$$h_2 = \frac{1}{2R^5}\{e^{-2kR}[(kR)^3 + 6(kR)^2 + 15kR + 18] -$$

$$- \frac{9}{2} I_{\frac{1}{2}}(kR) K_{\frac{1}{2}}(kR)\} \qquad (32)$$

($I_{\frac{1}{2}}(x)$ and $K_{\frac{1}{2}}(x)$ are modified Bessel functions of order $\frac{1}{2}$ of first and second kind, respectively) to see how again one has an oscillating term which may very easily predominate over the others, therefore giving a negative E_2. If that is the case, the spherical shape is unstable and the system will not have it as its equilibrium configuration.

To conclude, note a few general observations. Of course, as far as experimentalists are concerned, the problems to be faced are very difficult and the experiments critical, because quite small droplets should be observed, although bypassing the difficulties appears possible. As for the nuclear analogy, the same fact of having a non spherical shape generates rotational states for the droplet, and one has many dynamical analogies with nuclear theory: a number of concepts from nuclear physics could be applied to the present problem and, vice versa, useful indications could be brought from this approach into the consideration of nuclear matter. Indeed if one starts from a spherical droplet, and begins deforming it, e.g. giving it an increasing eccentricity along an axis, in general, if it was stable the energy will grow, but one can however, even in this case, find intermediate minima: this kind of result is very similar to that which Strutinsky[2] computed in fission theory, and had a significant impact on the theory of heavy nuclei.

REFERENCES

1. N. Bohr and J. A. Wheeler, Phys. Rev. $\underline{56}$, 426 (1939).

2. V. M. Strutinsky, Nucl. Phys. $\underline{A122}$, 1 (1968).
   V. M. Strutinsky and S. Bjørnholm, Nucl. Phys. $\underline{A136}$, 1 (1969).

3. For general reference see e.g.: I. M. Khalatnikov, "Introduction to the Theory of Superfluidity," W. A. Benjamin, Inc., New York, 1965; J. Wilks, "The Properties of Liquid and Solid Helium," Oxford University Press, London, 1967; W. E. Keller, "Helium 3 and Helium 4," Plenum Press, New York, 1969.

4. A. D. B. Woods and R. A. Cowley, Can. J. Phys. $\underline{49}$, 177 (1971).

5. J. Gavoret and P. Nozières, Ann. Phys. (N.Y.) $\underline{28}$, 349 (1964).

6. L. D. Landau, J. Phys. U.S.S.R. $\underline{5}$, 71 (1941).

7. A. Molinari and T. Regge, Phys. Rev. Letters $\underline{26}$, 1531 (1971).

8. H. Lamb, "Hydrodynamics," Cambridge University Press, London, 1932.

9. T. Regge, J. Low Temp. Phys. $\underline{9}$, 123 (1972).

10. See e.g.: A. L. Fetter and J. Walecka, "Quantum Theory of Many Particle Systems," McGraw-Hill, New York, 1971.

11. M Rasetti and T. Regge, to be published.

QUANTUM STRUCTURAL STABILITY AND BIOLOGY

M. Dresden

Institute for Theoretical Physics

State University of New York at Stony Brook

Stony Brook, New York

I. BACKGROUND, BIOLOGICAL UNIVERSALS, TIME EVOLUTION
A. Background

It is not at all obvious that the ideas and methods of physics, especially those of theoretical physics, can be effectively used to gain an understanding of biological phenomena. These methods relate principles of some degree of universality to a large number of special situations and so provide a unified understanding of a class of phenomena and the possibility of prediction. It is well to recall that these "general principles" result almost always from an organization and extrapolation of empirical regularities, cast in a mathematical form. The further mathematical analysis of these principles then (hopefully) leads to the specific results.

If a scheme of this general type is to be effective in biology, there must exist a number of universal elements in biology, which similarly allow the formulation of a set of principles, ideas and concepts of general scope and validity, which again at some stage can be put in mathematical form. It is obviously not necessary that these principles encompass all of biology, any more than regularities in physics have to possess complete universality in order to be useful. However a fairly broad class of phenomena should be contained within the description envisaged. It is precisely in this respect that the spectacular results of modern biology are most encouraging; in spite of the overwhelming overall complexity, a number of universal features are clearly discernable.[1]

Example 1. All cells have a universal structure, containing a cell wall, (a membrane), a replicating apparatus, an energy transfer mechanism. The replicating apparatus always contains nucleic acids, the energy transfer always proceeds via the formation or breaking of P bonds.

Example 2. All the complex biochemical reactions take place via a large number of individually simple steps. Each reaction is catalyzed by an enzyme. The total effect of an enzyme in a reaction can be summarized as a change of the reaction rate. Even relatively simple reactions pass through a number of yet simpler stages, each catalyzed by an enzyme. For example, the conversion of one sugar molecule into two lactic acid molecules $C_6H_{12}O_6 \rightleftarrows 2C_3H_6O_3$, requires (at least) 11 separate reactions each with its own enzyme.

Example 3. The structure and function of each cell is largely determined by the proteins it contains. There exists of course an enormous number of proteins. The structure and function of each protein is largely determined by the sequence of amino acids it contains. Each protein contains at most 20-30 distinct types of amino acids. The manufacture of proteins, in particular the construction of the appropriate sequence of amino acids is always organized and controlled by two nucleic acids (DNA and (or) RNA). Each nucleic acid contains just four bases.

Even these very few examples already indicate the large extent to which general biochemical features govern biological processes. It is particularly important to observe that the obviously very complicated and interlocking biological processes can be described in terms of a large number of much simpler processes each involving many fewer types of constituents. Thus the observed complexity is - at least to a considerable extent - due to the cumulative effect of an enormous number of simpler individual steps. It is this feature in particular which suggests that appropriate adaptations of the methods of statistical physics (which are specifically designed to describe the combined effect of large numbers of individual processes) might well be effective in describing interesting and meaningful aspects of biological phenomena.

B. Time Evolution

The present discussion will be restricted to questions of the time behavior of biological systems. In view of what has been noted before, it can be anticipated that modifications or adaptations of the methods of theoretical physics dealing with the temporal development will be especially appropriate for this discussion. Thus the use of various forms of rate equations may be anticipated; a general mathematical analysis of dynamical systems can be expected to be useful inasmuch

as these methods deal specifically with time evolution.
There exist a considerable number of studies in which
the biological developmental processes have been described in just these terms.[2,3] Other areas within
physics such as irreversible thermodynamics and non-
equilibrium statistical mechanics also describe the
temporal development of systems. It is therefore not
surprising that the concepts from these areas (especially the former) are currently extensively used in the
description of biological developmental processes.[4]

As already mentioned the present discussion will be
restricted to aspects of the time behavior of biological
systems. However this topic itself is immense, covering
as it does all growth processes, morphogenesis, evolution
and many others. The discussion therefore will be restricted further to two topics:

General aspects of morphogenesis - Morphogenesis
includes that varied group of time dependent processes
whereby an organism develops to a particular shape, size
or form. For the present purposes it is important to
stress that this developmental process is a highly
patterned process. This means that there are a number
of stages (regimes) within which the process proceeds
rather smoothly. However at fairly well defined times,
discontinuous changes occur. For example, a cell will
grow in a fairly smooth manner, although cell division
is a dramatic, discontinuous event, occuring at quite
specific intervals. The most characteristic and also
most general feature of morphogenesis is this patterned
development with distinct characteristics in the various
regimes and the sudden, discontinuous transitions between
regimes. It is further worth noting that the pattern
is quite stable. It is not at all easy to interrupt or

alter the pattern without killing the organism. (Strictly speaking such alterations are possible, however, substantial external influences are required to produce such changes.) For a more precise discussion, the notion of "neighboring" system is needed. See section III.

Comments on a recent study of Eigen[5] on the evolution of biological macromolecules - The question of just how biological materials evolved from simple compounds is of course one of great profundity and almost forbidding difficulty. The view adopted in Ref. 5 is that at a certain stage a chemical evolution process had produced a number (if not all) of the basic building blocks needed for the production of biological macromolecules. Thus the presence of appropriate amino acids of nucleoside phosphates such as ATP and many others is simply assumed. The central problem is then how, given these materials (and appropriate external conditions of temperature and other factors), these ingredients evolved into biological macromolecules capable of some basic biological functions such as self organization or self replication. It is clear that whatever mechanism is envisaged for such a development, a large number of random events will inevitably take place. Therefore the problem is to find out how certain types of molecular species - presumably formed by random events in a fluctuating environment - can be selectively amplified within this environment. This requires a formalization of the notions of selection and competition. Once these are defined the time evolution of such systems can be studied and the eventual dominance of one species or the extinction of another may be investigated. This approach of course does not demonstrate the inevitable development towards self organizing units of increasing order;

it does (if it were correct) show the possibility of some selected development.

C. Remarks

The first approach towards an understanding of these features is to see whether simple mathematical schemes, hopefully based on reasonable physical, chemical and biological principles, can be set up which will reproduce the types of behavior just described. The specific characteristics to be exhibited by the scheme are the patterned development in the case of morphogenesis; the macromolecular evolution requires the demonstration of a selective development of one molecular species over another with an accompanying increase in organization and order.

It is worth stressing that the main point must be the qualitative understanding of the time behavior of the mathematical models. The phenomena in both classes are quite general; thus the results can not depend sensitively on the quantitative details of the description. Qualitative studies of mathematical systems are of course not at all new.[6] So far these qualitative studies have been more common in mathematics and engineering than in physics or chemistry. It now appears that the qualitative methods also provide the proper tools for the investigation of these biological phenomena. Thus the kind of information which must be extracted from the mathematical description of the system is rather different from that in most parts of physics. Questions of the stability of states, systems, or patterns assume an increasing importance, while the concept of structural stability[7,8] plays a central role in all the mathematical models of biological systems. The concept will be defined and explained in section III; it is

enough to remark here that it is a global property of a
system, expressing the fact that a small variation in
the parameters describing the system does not change the
qualitative character of the system. Apart from having
very desirable mathematical properties, it would appear
from the persistence of biological entities in variable
environments that some such characteristic as structural stability must be an essential property of any
biological system.

Thom[9] in an interesting and suggestive book has
constructed a framework for morphogenesis based on a
deep mathematical structure analysis of structurally
stable systems. It would appear however that for actual
applications a number of additional assumptions and
identifications need to be made, so that it is not clear
that the essential character of the morphogenetic
processes is totally contained in the mathematical analysis. These questions are discussed in some detail in
section III.

It is clear that whatever particular model is
picked or whatever simulation is constructed the systems
considered must be open. Biological systems receive
energy, momentum, material from the environment; the
interaction with the surroundings is an integral part of
the growth or development process. Thus it must be anticipated that a description by means of equilibrium
thermodynamics is not appropriate; essential biological
processes such as diffusion are inherently dependent on
non-homogeneous spatial distributions, again showing the
essential role of the non equilibrium processes. That
the local biological processes tend to increase the
order (decrease the entropy) was first noted by Szilard[10]
stressed by Schrödinger[11] and has now found a more

mathematical formulation in the recent work of Prigogine and Glansdorff.[12] As was already mentioned, it is not surprising that the biological processes of open systems will require a nonequilibrium description. The stability questions will then be of a dynamic character, and fluctuations can be expected to play a dominant role.

Section II gives a more detailed discussion of the general type of equation underlying many biological models. A critical analysis of Eigen's equations and a series of conjectures necessary to validate the claims conclude this section. Section III includes a brief exposition of the ideas of Thom. A specific example is given and some concerns about the detailed applicability are expressed. Section IV deals with the remarkable absence of any quantum notions in all these considerations. The paper concludes with some speculations about the role of quantum concepts and quantum structural stability.

II. BIOCHEMICAL DETERMINISM, MATHEMATICAL MODELS
A. Biochemical Determinism

A very common way to describe biological processes is to assume that the biological state at time t is fully determined by a finite set of concentrations $c_i(t)$, i=1, ..., N. The simplest assumption is that the concentrations depend just on the time. The c_i could be the concentrations of amino acids, proteins, enzymes, repressors, but the complete set of concentrations is supposed to give a complete specification of the biochemical state and hence (by assumption) of the biological state. The dynamics is determined by chemical reaction rate theory which yields a system of first order equations for c_i:

$$\frac{dc_i}{dt} = X_i(c_1,\ldots,c_N,t), \quad i=1,\ldots,N \quad . \tag{1}$$

If the X_i satisfy certain regularity conditions, the initial values of c_i determine the solutions of (1) uniquely. The "explanation" of a phenomenon now amounts to a demonstration that the time development of the phenomenon in question can be described as a possible solution of such a set of equations.

The reaction mechanism is contained in the functions X_i whose form is determined by that mechanism. In most practical cases the X_i are polynomials in the concentrations; in almost all cases they are algebraic functions. Thus all the biochemical information is collected in the functions X_i. In general the equations are nonlinear; it is this circumstance which allows a large variety of possible time behaviors. There could be a single steady state which could be reached from any initial state, there could be a number of steady states, oscillatory behavior could occur, etc.; the detailed behavior depends on the detailed form of X. The mathematical understanding of systems of nonlinear ordinary differential equations is reasonably good. Computer studies tend to be effective so that the time evolution can be systematically studied. Within this scheme, the biochemical information is fed in via the form of X (if there are autocatalytic actions of material 1, X_1 must contain a power of c_1; if materials 1 and 2 inhibit each other X_1 and X_2 must contain coupling terms, etc.) and the mathematical analysis of the time behavior can be checked experimentally.

Although many studies follow the procedure outlined with considerable success, these models really omit many essential factors. In many situations external influences add driving terms to equation (1). Although this may affect the nature and type of the solutions to some considerable extent, it is conceptually at any rate not a major change. However in many actual circumstances the concentrations depend on a number of other variables, such as temperature, location, etc. In such cases the equations become partial differential equations of the general type

$$\frac{\partial c_i}{\partial t} = D_i \Delta_x c_i(t,x,\xi) + X_i(c_i,x,\xi,t), \quad i=1,\ldots,N. \quad (2)$$

Here the x variables are positions; the ξ variables are other parameters. As written (2) describes a diffusion process in addition to the chemical reaction. The D_i are the diffusion coefficients. Δ_x is the Laplacian. (The spatial variables x, are often included among the ξ parameters.) The point to note about the system (2) is that it is a nonlinear system of **partial** differential equations and the **general** theory is in much poorer shape than the corresponding theory of ordinary differential equations. It is also clear that (2) requires the specification of boundary conditions, e.g., some of the features (semipermeability) of a cell wall could in principle be incorporated as boundary conditions. (This is in principle **impossible** for the system (1).) Furthermore it is certainly realistic to recognize that the concentration of enzymes or repressors is certainly **not** spatially homogeneous, so that it is appropriate to investigate the dependence of the concentrations on the positions. This all shows that the system (2) is more

interesting than (1). However because of the absence
of a general theory the main effort has gone in the discussion of special examples (the original paper of
Turing[2] did involve position dependent concentrations)
and relatively few general features are known. The
study of (2) starts with an examination of the auxiliary
problem:

$$\frac{dc_i(\xi,t)}{dt} = X_i(c_1,\ldots,c_N,\xi,t), \quad i = 1,\ldots,N. \qquad (3)$$

This describes the time evolution of the concentration
system for fixed ξ. It is then reasonable to hope that
the diffusion terms of equations (2) are small corrections imposed on (3), so that the qualitative aspects
of the time evolution even of (2) are correctly described
by (3). This is indeed the view adopted in most applications (and in section III), but a general assessment
of the validity of this procedure (especially in cases
where the ξ parameters such as the temperature also are
position dependent) is still lacking.

B. The Eigen Equations

An interesting illustration of the method, philosophy (and dangers!) of the approach just described is
contained in a study by Eigen[5]. In this investigation
a set of rate equations for the formation and removal of
molecular species are obtained. Let c_j be the concentrations of species j at time t, then the equations
are

$$\frac{dc_j}{dt} = (F_j - R_j)c_j + \sum_{\ell \neq j=1}^{N} M_{j\ell} c_\ell, \quad j=1,\ldots,N. \qquad (4)$$

In (4) F_j represents the rate of formation and R_j the
rate of removal of species j. Both of these summarize

a number of fairly complicated processes. Both F_j and R_j <u>could</u> depend on the other concentrations. F_j includes the formation of species j by duplication from j. The $M_{j\ell}$ term in (4) represents the formation of j from ℓ, the result of a copying <u>error</u>, a copy of species ℓ becoming j by mistaken duplication. The effectiveness of this miscopying is measured by the "mistake matrix" M.

For given F_j, R_j, M (all of which could depend on the c's)(4) is a standard type rate equation (see (3)); the question is really what kind of information is expected from it. This is not just a solution of the initial value problem, instead certain constraints and restrictions are imposed which incorporate or simulate some of the actual conditions under which the formation and removal of molecules in a cell take place. Assume first that F_j has the form

$$F_j = k_o a_j q_j . \qquad (5)$$

Here k_o is a rate constant, and a_j measures the <u>ideal</u> duplication rate. $1-q_j$ is a measure of the copying errors; if $q_j=1$, there are no errors. From this <u>interpretation</u> and the interpretation of M it follows that

$$k_o \sum_{j=1}^{N} c_j a_j (1-q_j) = \sum_{j \neq \ell = 1}^{N} M_{j\ell} c_\ell . \qquad (6)$$

(Both sides of (6) yield the total rate of miscopying.) While (5) is an assumption which can be made, (6) is a constraint <u>demanded</u> by the interpretation. The form for R_j is assumed to be

$$R_j = k_o D_j + \Phi_o / \Sigma c_i \qquad (7)$$

STRUCTURAL STABILITY

Pictorially the removal processes are represented by a diffusion type term characterized by D_j and a general dilution term due to the influx of material characterized by Φ_o. Obviously (5) and (7) represent special but reasonable choices for F_j and R_j. These choices can be substituted in (4) yielding (<u>without</u> using the constraint equation (7))

$$\frac{dc_j}{dt} = k_o W_j c_j - c_j \frac{\Phi_o}{\Sigma c_i} + \sum_{\ell \neq j=1}^{N} M_{j\ell} c_\ell \, , \quad j = 1, \ldots, N, \quad (8)$$

$$W_j \equiv a_j q_j - D_j \, , \quad (8a)$$

where W_j is called <u>selective advantage.</u>

Thus the system (8) is simply a specialized form of the basic equation (4). At this point an additional condition is imposed expressing the biological circumstance that the <u>net</u> total production rate shall be zero. Using the formulae (5), (7) and the constraint (6) yields a specific expression for the dilution rate Φ_o:

$$\Phi_o = k_o \sum_{j=1}^{N} c_j (a_j - D_j) \, . \quad (9)$$

The equations (8) must be discussed subject to the constraints (9) and (6). Substituting (9) in (8) automatically incorporates (9), giving

$$\frac{dc_j}{dt} = k_o c_j \left(W_j - \frac{\Sigma c_i \varepsilon_i}{\Sigma c_i} \right) + \sum_{\ell \neq j=1}^{N} M_{j\ell} c_\ell \, , \quad j=1,\ldots,N, \quad (10)$$

where

$$\varepsilon_i \equiv a_i - D_i \, . \quad (10a)$$

It is easy to show from (10), (10a) and (8a) using (6) explicitly that

$$\frac{d}{dt} \Sigma c_j = 0 \quad \text{or} \quad \Sigma c_j \equiv n = \text{constant}. \tag{11}$$

Thus the constraint (6) and the condition that the net production rate shall vanish can be combined into the requirement that the total concentration is fixed. This can also be inverted to observe that (10) together with the definitions for W and ε and (11) imply the constraint equation (6). It is therefore sufficient to study the system (10) just with the constraint (11). It is this overall conservation law which introduces the competition between the species. A gain for one must come at the expense of another. It should be noted that since the operations leading to (10) are purely algebraic, it is perfectly legitimate for a_j (in (5)) and D_j (in (7)) to depend on the concentrations. This in turn will produce a (known) c dependence in W_j and ε_j.

The basic question can now be phrased in this fashion: the various species are characterized (and differentiated by) the numerical values of W_i and ε_i. The time evolution of the interacting competing system is described by (10) and (11). Is it true that for a given initial situation, one or a number of species evolve as the dominant ones? If so, what in terms of W_i and ε_i are the characteristics of the selected species?

For the mathematical discussion it is more convenient to absorb k_o in W_j and ε_i, and to use $\frac{c_j}{n}$ as new variables, the basic equations become then

$$\frac{dc_j}{dt} = c_j(W_j - \sum_{i=1}^{N} \varepsilon_i c_i) + \sum_{\ell \neq j=1}^{N} M_{j\ell} c_\ell, \quad j=1,\ldots,N, \tag{12}$$

with

$$\sum_{j=1}^{N} c_j = 1. \tag{12a}$$

STRUCTURAL STABILITY 369

The further interpretations of these equations as those of a competing system, describing a selection process, rests on certain mathematical conjectures. These can be phrased in this way:

If the W_j, ε_j, M_{ij} are all independent of the concentrations c_j, there exists a single species s, such that for any initial configuration

$$\lim_{t \to \infty} c_s(t) = 1 \quad , \tag{13a}$$

$$\lim_{t \to \infty} c_r(t) = 0 \quad , \quad r \neq s \quad . \tag{13b}$$

If the $W_j, \varepsilon_j, M_{ij}$ depend on the concentrations c in a smooth manner (allowing a Taylor development) there exists a set of ℓ species, $\ell \ll N$, such that for any initial configuration and for any i within the set $1, \ldots, \ell$,

$$\lim_{t \to \infty} c_i(t) = \bar{c}_i, \quad i \in \{\ell\} \quad , \tag{14a}$$

$$\lim_{t \to \infty} c_j(t) = 0 \quad , \quad j \notin \{\ell\} \quad . \tag{14b}$$

The limiting values $\bar{c}_i$ will in general depend on the initial data. There will further exist a set of real numbers $v_1, \ldots, v_\ell$, such that

$$\bar{c}_1^{v_1} \ldots \bar{c}_\ell^{v_\ell} = \text{constant} \quad . \tag{14c}$$

If these conjectures are correct, they express the evolution to dominance of a single species or a small set of species and the extinction of all others. Alternately these results give a formal expression to "survival of the fittest". (14c) expresses a type of chemical equilibrium between the surviving interacting species. On the basis of these conjectures the equations (12) do provide a formalization of the selection notion. Unfortunately, although the equations do describe some such process, the conjectures as phrased are not correct; especially, (13b) is wrong, no species really dies out. This can be seen explicitly in (12) which can be rigorously solved. Define first

$$\bar{\varepsilon} = \sum_{i=1}^{N} c_i \varepsilon_i \qquad . \qquad (15)$$

(Notice $\bar{\varepsilon}$ is <u>not</u> known.) Now introduce the new variables

$$z_i(t) = c_i(t) \exp [\int_0^t dt' \bar{\varepsilon}(t')] \quad . \qquad (16)$$

Clearly one has

$$z_i(t)/z_j(t) = c_i(t)/c_j(t) \quad . \qquad (16a)$$

Substitution of (16) in (12) yields a linear equation for the z variables

$$\frac{dz_i}{dt} = \sum_{\ell=1}^{N} Q_{i\ell} z_\ell \, , \, Q_{i\ell} = M_{i\ell}, \, i \neq \ell; \, Q_{ii} = W_i \quad . \qquad (17)$$

Since M and W were independent of c, Q is a constant matrix and (17) can be directly solved in terms of the eigenvalues $\lambda_1, \ldots, \lambda_N$ of Q. (It will be assumed that these eigenvalues are all distinct, the necessary modifications in the case this is false are straightforward.) Since Q is (by assumption) a real symmetric matrix, it can be diagonalized by a unitary matrix U. In terms of U the solutions for z can be written in the form

$$z_i(t) = \sum_{j,\ell} C_{ij\ell} \exp(\lambda_j t) z_\ell(0), \quad i=1,\ldots,N, \qquad (18)$$

where

$$C_{ij\ell} = U_{ij} U_{j\ell}^{-1}, \quad i,j = 1,\ldots,N. \qquad (18a)$$

In (18a), there is <u>no</u> summation over j, as of course

$$\sum_j C_{ij\ell} = \sum_j U_{ij} U_{j\ell}^{-1} = \delta_{i\ell}.$$

(18) shows the role of the initial data, the time evolution and the properties of the system. Let λ_m be the maximum eigenvalue; then it follows directly from (18) that the large time $(t\to\infty)$ behavior is controlled by that largest eigenvalue. If i and j are fixed and such that $C_{im\ell}\neq 0$ and $C_{jm\ell}\neq 0$ for some ℓ, then it follows from (18) and (16a) that

$$\lim_{t\to\infty} c_i(t)/c_j(t) = [\sum_\ell C_{im\ell} z_\ell(0)]/[\sum_\ell C_{jm\ell} z_\ell(0)]. \qquad (19)$$

The property of possessing some nonvanishing matrix elements $C_{im\ell}$ depends on the diagonalizing matrix U, thus on Q but <u>not</u> on the initial data. (19) clearly shows that <u>no</u> species really dies out; the concentration ratios remain finite; (13b) is not satisfied. If initially one species is present $z_1(0)=1$, $z_i(0)=0$, $i\neq 1$, the same formulae yield

$$c_i(t)/c_j(t) = [\sum_\ell C_{i\ell 1} \exp(\lambda_\ell t)]/[\sum_\ell C_{j\ell 1} \exp(\lambda_\ell t)]. \qquad (20)$$

The time behavior of this ratio depends on the relation between the largest λ_ℓ for which $C_{i\ell 1}\neq 0$, and the largest λ_ℓ for which $C_{j\ell 1}\neq 0$. If these are called λ_a and λ_b respectively there results

$$\lim_{t \to \infty} c_i(t)/c_j(t) \to A\, e^{(\lambda_a - \lambda_b)t}, \quad A = \text{constant}. \tag{21}$$

This will obviously decay, grow or remain constant depending on the relative sizes, but again this is controlled by the details of the U matrix. No simple general conclusion about the growth or extinction independent of the nature of the system exists. The details of the Q matrix influence the time behavior in an essential manner. The time behavior is <u>independent</u> of the ε_i variables. Nothing rigorous seems to be known about the situation where ε_i and W_j depend on c, so that the Eigen equations, in spite of their suggestive features, do not seem to be a satisfactory model for competition and selection.

III. ARE THERE UNIVERSAL MATHEMATICAL MECHANISMS?
A. Structural Stability

Although the examples given are all suggestive of a certain kind of biological behavior, no general mathematical mechanism has been proposed. The phenomena were described for particular classes of equations, with special expressions for the X_i. Furthermore, in a realistic situation the functions X_i in the rate equations are often incompletely or only approximately known; it is not clear how much significance can be attributed to detailed results derived on the basis of particular expressions. To solve the equations initial data are also needed and these are often known to a limited accuracy. On these grounds there might be considerable doubt about the effectiveness of the procedures discussed. What is needed is a method to describe the <u>typical</u> behavior of a

certain _type_ of system. The typical behavior refers to
the general qualitative _character_ of the process; the
type of system refers to the general _features_ of the
system. These ideas can be made precise by using the
concept of structural stability. Consider the basic
form of rate equations (3). For fixed ξ, (3) is a set of
first order differential equations which define a unique
trajectory, $c_i(t)$, i=1,...,N, through each point of the
M space, a point in M being denoted by $c \equiv (c_1,...,c_N)$.
The initial data needed to fix a solution of (3),

$$c_i(0) = b_i, \quad i = 1,...,N, \qquad (22)$$

specify that b is a point on the trajectory $c_i(t,b)$.
The totality of _all_ the trajectories on M is called the
phase space pattern (phase space portrait). The phase
space pattern on M gives all possible orbits of a given
dynamical system X, for all possible initial data.

It is clear from (3), that the $X_i(c,\xi)$ may be viewed
as the components of a vector at the point c in M tangent
to the trajectory through c. X therefore defines a
vector field on M. The advantage of considering X as a
vector field on M is that now the notion of a _neighboring_
vector field on M can be defined in a natural manner.
X and Y are neighboring vector fields if the components
are everywhere close as expressed by

$$|X_i(c) - Y_i(c)| < \varepsilon \quad \forall \, i, c. \qquad (23)$$

If ε is independent of c, (23) defines a particular
topology. Since each vector field X on M defines a
dynamical system, the notion of neighboring dynamical
system is now direct. In the situation just described,

there are two dynamical systems, X and Y, on M. Each
of these systems has of course its own family of trajectories in M; each has its own phase space pattern.
Two systems are <u>equivalent</u> if their two phase space
patterns are <u>homeomorphic</u>. A system X is structurally
stable if there is a neighborhood of X such that for all
systems Y in the neighborhood $|X-Y| < \varepsilon$ the systems X
and Y are equivalent. The phase space patterns for a
structurally stable system are homeomorphic for a class
of neighboring vector fields. Since for structurally
stable systems the qualitative aspects do not change
from one set of X_i to a neighbor it is not of great
concern that the X_i may not be known with complete precision. Furthermore the phase space patterns describes
the qualitative behavior of the system for <u>all</u> initial
data; that the initial data are only partially known is
not of any great consequence. Thus it seems that the use
of structurally stable systems is natural for biological
systems.

<u>Remark 1.</u> In many examples the rate equations form
a gradient system; there is a function V (with appropriate regularity properties) such that

$$X_i = -\frac{\partial V}{\partial c_i} \quad . \tag{24}$$

The neighborhood is expressable in terms of potentials:
one could for example define V near V' if

$$|V(c)-V'(c)| < \varepsilon \quad \forall c \in M \quad . \tag{25}$$

This would be the C^0 topology; but one could in addition
demand the proximity of partial derivatives of V to any
desired order. The C^1 topology would require in addition

STRUCTURAL STABILITY

to (25) that

$$\left|\frac{\partial V}{\partial c_i} - \frac{\partial V'}{\partial c_i}\right| < \varepsilon \quad \forall\, i,\ \forall\, c \equiv M. \tag{26}$$

There is no a priori reason to prefer one topology over another.

Remark 2. Typical systems in physics are not structurally stable. Consider a harmonic oscillator described by

$$\frac{dx}{dt} = p, \quad \frac{dp}{dt} = -kx, \tag{27}$$

where p is the momentum, x is the position, and k is a constant. Its phase space pattern is a family of closed curves (ellipses). Consider the neighboring system defined by

$$\frac{dx}{dt} = p, \quad \frac{dp}{dt} = -kx + \varepsilon p. \tag{28}$$

For arbitrary small ε the space pattern is a family of spirals, spiralling into the origin. Thus the phase space patterns of the neighboring system and the harmonic oscillator are _not_ homeomorphic; the oscillator is _not_ structurally stable.

In general Hamiltonian systems are _not_ structurally stable.

B. Conjectures About Dynamical Systems

The description given allows the comparison of vector fields or dynamical systems on a space M (usually assumed to have certain smoothness properties). Consider the space D of dynamical systems on M, each element of D being a vector field on M. D is a vector space. A topology is defined since the notion of neighboring dynamical system is specified via (25), (26) or (23). Call the subset of D which consists of structurally stable systems S, its complement Σ, the <u>bifurcation set</u>. The boundary set separating the set of structurally stable systems S from Σ is the <u>catastrophe set</u>. Consider a system defined by $X_i(c,\xi)$: for a given ξ there will be a given system corresponding to a point in the space D. Assume that this point lies in S. As ξ varies the point in D will wander. As long as it stays in S the qualitative structure will not change. However as this point enters Σ there will be a radical qualitative change in the system's behavior. If S and Σ have a layered structure in D variations of ξ will give rise to a series of smooth structurally stable regimes separated by catastrophes. The geometry of the space D and its eventual splitting in strata corresponding to S and Σ are mathematical questions. It is very suggestive to identify the various regimes in the morphogenetic process with the passage of the representative system point through the various strata in the space D. This is the basic idea of the Thom theory. It should be stressed that what is described here refers to motion in the D space the space of dynamical systems, not in the phase space (M space) of the system.

It should be finally noted that the nature of the space D and the set S crucially depend on the topology imposed. For example a result of Peixoto asserts that if the space M is two dimensional and compact, S is dense in D. (This employs the C^1 topology.) If the dimension of M is larger than 2, then the set S is not dense in the C^1 topology. Obviously the general study of D and S are problems of considerable difficulty.

It is worth recalling that the actual behavior of the dynamical system is described by a single trajectory in the space M. The understanding of the phase space pattern is facilitated by the use of the <u>attractor</u> notion. A region A of M is an attractor with domain B(A), if any orbit passing through a point in B(A) approaches A as t→∞. The idea of an attractor is a generalization of that of a limit point. For simple systems the attractors are points or limit cycles. In more general cases they can have quite complicated topological structure. The corresponding domains can be interwined in a most complicated fashion and a slight change from one point to another can shift a trajectory from the domain of A to that of another attractor A'. If a system depends on parameters ξ, it is important to study the dependence of the nature and number of the attractors on these parameters. Since the attractors describe the phase space pattern information about them is important. Most is known about <u>potential</u> systems; about general systems much less is known but the following properties are conjectured:

First, for almost all vector fields on a smooth M almost all trajectories tend rapidly toward attractors.

Second, most systems have a finite set of structurally stable attractors $A_1, \ldots, A_s$ with domains $B(A_i)$.

Third, $$M = \bigcup_{i=1}^{s} B(A_i) \, .$$

Fourth, the surfaces separating the domains are structurally stable as well as differentiable.

C. An Example

To get some feeling for the attractor structure and their changes as the parameters change, consider a very simple case where M is a 2-dimensional Euclidean space. The two concentrations c_1 and c_2 satisfy

$$\frac{dc_1}{dt} = c_1\xi_2 - c_2 + c_1\xi_1\sqrt{c_1^2+c_2^2} - c_1(c_1^2+c_2^2)$$
$$\frac{dc_2}{dt} = c_2\xi_2 + c_1 + c_2\xi_1\sqrt{c_1^2+c_2^2} - c_2(c_1^2+c_2^2)$$
(29)

Both c_1 and c_2 are deviations from standard concentrations, hence they can be positive and negative. ξ_1 and ξ_2 are real parameters. (29) can be simplified by introducing polar coordinates giving

$$\frac{dr}{dt} = -r^3 + \xi_1 r^2 + \xi_2 r \qquad (30a)$$

$$\frac{d\theta}{dt} = 1 \qquad (30b)$$

The sign of $\frac{dr}{dt}$ (expanding or contracting spiral) follows directly from (30a); it is clear that there are at most three circular orbits, call

$$r_\pm = \frac{1}{2}\xi_1 \pm \frac{1}{2}\sqrt{\xi_1^2 + 4\xi_2} \qquad (31)$$

Now the ξ_1-ξ_2 plane (which is the parameter space or Ξ space) splits up in four regions (see Fig. 1), each corresponding to a distinct attractor structure in the M space. The change in the attractor structure with the variation of the parameters is an example of <u>bifurcation</u> theory.

Region I is that part of the Ξ plane where $\xi_1 > 0$ and $\xi_2 < 0$, between the parabola $\xi_1^2 + 4\xi_2 = 0$ and the ξ_1 axis. Corresponding to every point in this region is a dynamical system possessing <u>two</u> attractors, a limit cycle of radius r_+, and one point attractor at the origin[6]. The domain $B(r_+)$ is the region in the M space defined by $r_- < r < \infty$. The domain $B(0)$ is $r < r_-$.

Figure 1

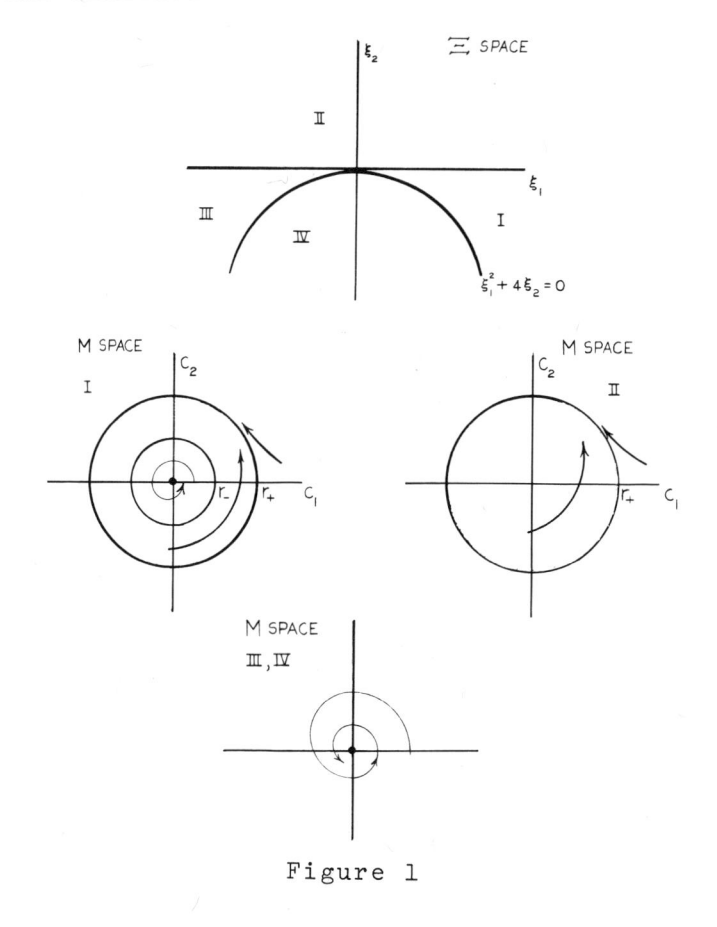

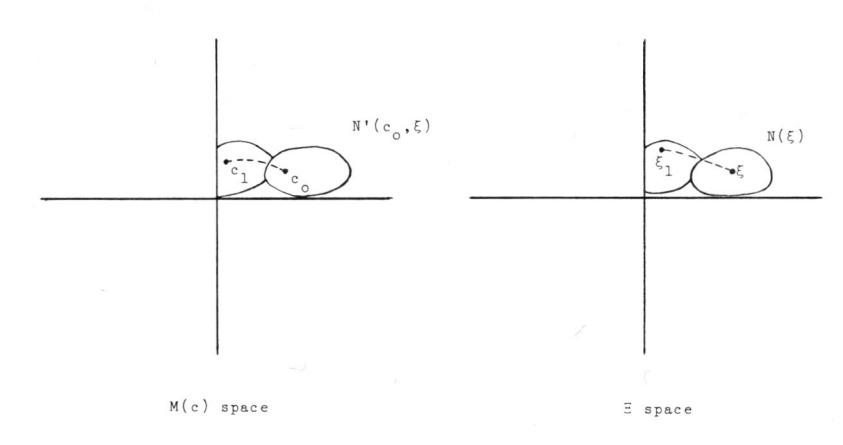

Figure 2

Region II is the upper half Ξ plane. Corresponding to every point in this region is a dynamical system possessing just <u>one</u> attractor, the circle of radius r_+. The domain of this attractor is the complete M plane. If the system starts at $r=0$, it stays there; for all other orbits r approaches r_+.

Region III is the part of the Ξ plane where $\xi_1<0$, $\xi_2<0$ between the parabola $\xi_1^2+4\xi_2 = 0$ and the ξ_1 axis.

Region IV lies below the parabola $\xi_1^2+4\xi_2=0$. For regions III and IV the dynamical system has one attractor, the origin. This example demonstrates quite clearly that a change in parameters alters the attractor structure quite drastically. As a consequence a slow smooth change in the ξ variables can cause a very radical alteration in behavior, for example, if ξ_2 changes from $\xi_2>0$ to $\xi_2<0$. As long as the ξ variables do not cross the boundaries the system changes smoothly with these variables. If the ξ variables do cross the boundary there is a sudden change again suggesting a connection between the bifurcation phenomena and morphogenesis.

D. Interpretation Assumptions of Thom Theory

This example illustrates the general features which will now be summarized. Consider again the general problem (3). The set of ξ values form the parameter space, Ξ. A dynamical system is associated with each point in Ξ space. The dynamical system is characterized by a set of vectors $X_i(c,\xi)$ or a potential $V(c,\xi)$ defined on M. The dynamical system is described by a phase space pattern in M. Since the main interest here is in the description of dynamical processes, it is more suggestive to stress the association of a complete dynamical system on M with each point in Ξ.

This dynamical system on M consists of trajectories, an attractor structure A, and a set of domains B(A). As ξ changes there may be changes in the attractor structure as illustrated in the last section. Thus the Ξ space splits up in a number of regions: as the parameters cross from one region to another discontinuous changes in the behavior of the system will occur. This mathematical phenomenon is <u>assumed</u> to be the underlying mechanism describing the characteristic patterned development of morphogenesis. When ξ does not cross the boundary (catastrophe set) the behavior of the system changes smoothly; otherwise the attractor structure on M and the qualitative behavior of the system change discontinuously. The interplay between the continuous smooth changes and discontinuous dramatic changes is a very suggestive description of the developmental processes.

For the actual applications, the picture as outlined here needs to be supplemented by a number of further interpretative assumptions. First, what needs to be described ultimately is something about the time behavior of a single trajectory in the M space. Within the picture proposed it is assumed that the system moves rapidly to a structurally stable attractor and stays there. When the attractor structure changes because of the <u>slow</u> changes in ξ, the system moves rapidly to a new attractor, so providing the evolutionary pattern. This description really requires a definition of slow and fast, thus one has a number of time <u>scales</u> in mind. It is essential that these time scales are widely separated: the system in the M space is assumed to reach a stable attractor in time intervals during which the variation in ξ space can be neglected. That is a rather strong assumption and nothing is known about its validity.

What is needed is the bifurcation theory of partial differential equations such as (2), but not much seems to be known about this in general.

Second, the reason that this lack has not hampered the use of these ideas is that a most important additional assumption is always made. Consider a gradient system where $V(c,\xi)$ defines the vectors X_i via (24). A single point in Ξ space corresponds to a complete dynamical <u>system</u> described by $V(c,\xi)$ evolving in time in M. The basic new assumption made is that the dynamical system can be described by a <u>single state</u> of that system. The time evolution of the system is assumed to be completed; customarily the state of the system is taken to be that of a stable attractor. In the case of a gradient system this state would be characterized by

$$\frac{\partial V(c,\xi)}{\partial c_i} = 0, \quad i = 1,\ldots,N \quad . \tag{32}$$

Thus <u>the</u> state singled out in the M space is the state of stable equilibrium at a single point $c = c_o$ which is a solution of (32); thus <u>the</u> state of the system corresponding to a fixed value ξ is a critical point, c_o, of the vector field V on M. For neighboring values of ξ there will be neighboring potentials on M with neighboring critical points near c_o.

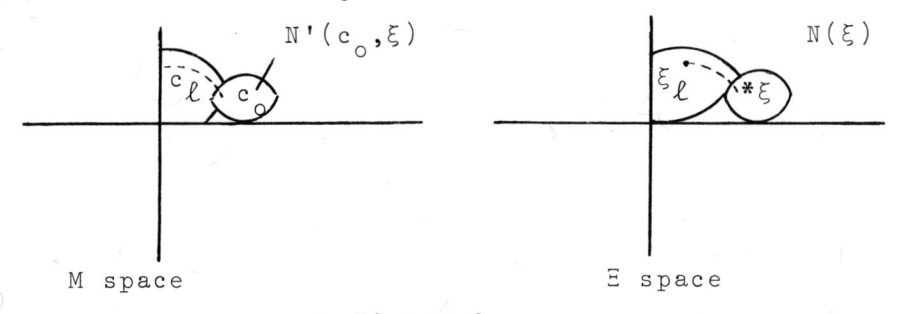

Figure 3

Let c_o be the state in M corresponding to ξ in Ξ. If the potential V is structurally stable there will be a neighborhood $N(\xi)$, of ξ in Ξ so that the states in M corresponding to $\xi \in N$ will lie in a neighborhood N' of c_o in M. Furthermore all these critical points of the vector field $V(c,\xi)$ on M will be of the <u>same type</u> as c_o. The neighborhood $N'(c_o)$ is the <u>domain</u> of the attractor which in this case is the singular point c_o.

Third, the theorem of Thom[9] allows a systematic and exhaustive enumeration of all the types of critical points for a Ξ space of given dimensionality. As the parameters are varied ξ may move out of $N(\xi)$; the corresponding critical point of the system then moves out of $N'(c_o,\xi)$ to a new domain belonging to c_ℓ. The character of the singular point c_ℓ is different from that of c_o. As the boundary is crossed in M a catastrophe occurs; because of Thom's theorem there is a well defined structural change.

In this manner a relationship is established between the regions in Ξ space and the domains of singular points in M space, the phase space of the system. That the limited number of morphological types is connected with the limited number of types of singular points is certainly most suggestive. However all this is accomplished at the cost of describing a dynamical system in terms of a single time-independent state. This makes it impossible to obtain <u>any</u> information about the time evolution of the system. Furthermore this also means that the resulting description depends crucially on the criteria chosen to obtain that single representative state. In most physical and biological systems the choice of the static equilibrium point (as done by Thom) seems arbitrary and inappropriate. For systems with concentration and temperature gradients this is certainly incorrect. Finally most of the results - especially the classification theorem - are apparently established only for gradient systems. The main systems of interest (those which could exhibit structural stability), however are not of that type. So, although suggestive, it cannot be claimed that the mathematical mechanism proposed is the only or the key ingredient in the explanation of the evolutionary processes.

IV. THE ROLE OF QUANTUM NOTIONS
A. General Remarks - Relevance

It is a remarkable fact that in none of the studies discussed do quantum notions appear at any stage. In a sense this is clear once the rate equations are written; all of the physics, chemistry, and, in fact, the complete underlying mechanism is contained in the functions and parameters occuring in the equations. Thus there might be a quantum theory of a diffusion coefficient, but once that theory is established the diffusion coefficient enters the rate equations as a classical parameter and its quantum origin is not relevant for the further discussion. The various ideas and concepts introduced, e.g., structural stability attractors, had their origin in classical dynamics and these notions are accordingly appropriate to classical situations.

The fundamental question is whether in the biological developmental process there are intrinsic quantum mechanical features, whether the time development in non-equilibrium quantum statistical mechanics can be phrased in a fashion that is mathematically similar to the classical version, not only with respect to the formulation, but also with respect to the <u>kind</u> of questions asked. The parameters in the two descriptions would be different. Ultimately this goes back to the phenomena itself; are there characteristic quantum mechanical time developmental features? The answer to this question is certainly yes: spin-echo and AC Josephson junction experiments are obvious examples. Are these characteristic quantum developmental features relevant for biological evolutionary processes? Since all these processes are intrinsically <u>macroscopic</u> in character it would seem that quantum phenomena cannot play a major role. Indeed it is hard to believe that quantum mechanics is of importance in the evolution of an elephant embryo. However the situation is quite different in the molecular evolutionary processes described by Eigen (section II). Quantum mechanics is essential for the description of molecular properties; so significant aspects of the formation process of new molecular species may be quantum mechanical in character, in which case the rate equations would have to be quantum mechanical and the discussion modified accordingly. That the masses of the objects involved are too large to necessitate quantum notions is not a convincing argument. There are certain features of large molecules which are definitely quantum mechanical in origin and there do exist macroscopic quantum phenomena. Macroscopic type quantum phenomena may not play a role in biological developmental processes, but the <u>possibility</u> has not been excluded.

B. Quantum Structural Stability

There does not appear to exist an obvious definition of structural stability which is appropriate

for quantum systems. Since it was stressed several
times that the idea of structural stability is useful
and natural in biological situations, it is interesting
to inquire about possible quantum modifications. The
main difficulty is that classical Hamiltonian systems
are never structurally stable and that quantum mechanics
strongly depends on the Hamiltonian formulation. One
can however still define structural stability. Consider
a system having a Hamiltonian H. Let $<x(t)>_\psi$ and
$<p(t)>_\psi$ be the quantum averages of position and momentum
at time t if the initial state is ψ. These averages
define a curve (for each ψ) in the $<x>$-$<p>$ space. The
quantum phase space pattern would be obtained by letting
ψ run through all admissable initial wave functions.
Two systems are equivalent if their phase space patterns
are homeomorphic; a system H is structurally stable if
there is a neighborhood N(H), such that all systems in
N(H) are equivalent to H.

In order to arrive at a somewhat general discussion
of these questions, it is necessary to use a non-
Hamiltonian form of quantum mechanics which allows the
description of explicitly time-dependent phenomena.
The most suitable formalism for that purpose is Feynman's
path integral method, based on a Lagrangian rather than
a Hamiltonian formulation. This program has been carried
out for the quantum damped harmonic oscillator, which
turned out to be structurally stable, while the undamped
harmonic oscillator is not. The definition of structural
stability used was the one given above. These results
only indicate that it is possible to define the quantum
structural stability notion, but its utility, relevance
and eventual significance remain to be investigated.

C. Time Scales

If it is indeed necessary to employ a quantum description of the evolutionary process the rate equations must be quantum mechanical equations. There is an extensive general theory of such equations but there are relatively few cases where exact solutions have been obtained which can be compared with the corresponding exact classical solutions. Only in this way can the specific quantum mechanical time evolution features be isolated. In a recent study[13] it was shown for highly idealized schematic systems that such features exist. The particular differences discovered were: an oscillatory approach to an equilibrium state in the quantum case, as contrasted with a monotonic approach to equilibrium in the classical case; a persistent correlation in the quantum case, as contrasted with a disappearance of correlations in the quantum case.

It was recently shown by Kenkre[14] that these features are not restricted to the very special systems considered in Ref. 13, but they are a fairly general consequence of quantum mechanical rate equations. Furthermore in these equations a series of time scales with quite distinct behaviors occur in a more or less natural manner. It would be most interesting if these ideas could be combined with the geometric approach to give a general method as well as a detailed understanding of evolutionary processes.

ACKNOWLEDGMENTS

The author would like to thank Professor E.C. Zeeman and Professor D.H. Fowler of the University of Warwick, and Professor Klaus Jänich of the University of Regensburg for their patient explanations of catastrophe theory to an unwilling, skeptical listener. It is hoped that the results of their teachings will not cause them to stop these efforts altogether.

REFERENCES

1. Green and Goldberger, Molecular insights into the living process, Academic Press, New York, 1967.
2. A.M. Turing, Phil. Trans. Royal Soc. B **237**, 5 (1952).
3. D. Ruelle, Trans. N.Y. Ac. Sci., **35**, 66 (1973).
4. I. Prigogine and G. Nicolis, Quart. Rev. Biophys. **4**, 107 (1971).
5. M. Eigen, Naturwis. **58**, 466 (1971).
6. V.V. Nemytskii and V.V. Stepanov, Qualitative theory of differential equations, Princeton University Press, Princeton, 1960.
7. R. Rosen, Dynamical systems theory in biology, Wiley, New York, 1970.
8. H.F. de Baggis in: Contributions to the theory of nonlinear oscillations, S. Lefschets et al. Eds., Princeton University Press, Princeton, 1950.
9. R. Thom, Stabilité structurelle et morphogénèse, W.A. Benjamin, Inc., Reading, Mass., 1972.
10. L. Szilard, Zs. für Physik, **53**, 840 (1929).
11. E. Schrödinger, What is life, Cambridge University Press, Cambridge, 1944.
12. P. Glausdorff and I. Prigogine, Thermodynamic theory of structure, stability and fluctuations, Wiley, New York, 1971.
13. M. Dresden and F. Feiock, Stat. Phys., **4**, 111 (1972).
14. Kenkre, private communication.

LIST OF PARTICIPANTS

Alexander A. Abela
Air Force Office of
  Scientific Research
Arlington, Virginia

Matti S. Al-Aish
National Institute of
  General Medical Sciences
National Institutes of
  Health
Bethesda, Maryland

James S. Ball
Department of Physics
University of Utah

J. Norman Bardsley
Department of Physics
University of Pittsburgh

Benjamin Bederson
Department of Physics
New York University

Lawrence C. Biedenharn
Department of Physics
Duke University

D. I. Blokhintsev
Laboratory of Theoretical
  Physics
Joint Institute for
  Nuclear Research
Dubna, U. S. S. R.

Fritz Bopp
Sektion Physik der Ludwig-
  Maximilians-Universität
Munich, Germany

Gregory Breit
Department of Physics
  and Astronomy
State University of New
  York at Buffalo

Steven Brown
Center for Theoretical Studies
University of Miami

Peter Carruthers
Department of Physics
Cornell University

John M. Charap
Queen Mary College
University of London
London, England

Mou-Shan Chen
Center for Theoretical Studies
University of Miami

Chinn-Chann Chiang
Center for Particle Theory
University of Texas at Austin

Fritz Coester
Argonne National Laboratory
Argonne, Illinois

Sidney Coleman
Department of Physics
Harvard University

T. Patrick Coleman
Center for Theoretical Studies
University of Miami

Anthony P. Colleraine
Department of Physics
Florida State University

Sorin Comorosan
J. M. Richards Laboratory
Grosse Point Park, Michigan

Fred Cooper
Belfer Graduate School
  of Science
Yeshiva University

LIST OF PARTICIPANTS

Horace Crater
Department of Physics
Vanderbilt University

P. A. M. Dirac
Department of Physics
Florida State University
    and
Center for Theoretical
  Studies
University of Miami

Max Dresden
Institute of Theoretical
  Physics
State University of New
  York at Stony Brook

Edward Eisenstein
Department of Biophysics
Michigan State University

Werner Eissner
Center for Theoretical
  Studies
University of Miami

P. S. Farago
Department of Physics
University of Edinburgh
Scotland

Bernard T. Feld
Laboratory for Nuclear
  Science
Massachusetts Institute
  of Technology

Ephraim Fischbach
Department of Physics
Purdue University

Wade L. Fite
Department of Physics
University of Pittsburgh

Ronald Fox
School of Physics
Georgia Institute of
  Technology

Sidney Fox
Institute for Molecular
  and Cellular Evolution
University of Miami

Peter G. O. Freund
Enrico Fermi Institute
University of Chicago

Herbert Friedman
U. S. Naval Research
  Laboratory
Washington, D. C.

L. W. Frommhold
Department of Physics
University of Texas at Austin

Edward Gerjuoy
Department of Physics
University of Pittsburgh

Sheldon Glashow
Department of Physics
Harvard University

Edwin L. Goldwasser
National Accelerator
  Laboratory
Batavia, Illinois

O. W. Greenberg
Department of Physics
  and Astronomy
University of Maryland

Martin B. Halpern
Department of Physics
University of California
  at Berkeley

Morton Hamermesh
School of Physics
  and Astronomy
University of Minnesota

Ronald J. Henry
Department of Physics and
  Astronomy
Louisiana State University
  at Baton Rouge

LIST OF PARTICIPANTS

Arvid Herzenberg
Department of Engineering
  and Applied Science
Yale University

Robert Hofstadter
Department of Physics
Stanford University

Joseph B. Hubbard
Center for Theoretical
  Studies
University of Miami

Hans A. Kastrup
Institut für Theoretische
  Physik der Rheinisch-
  Westfälischen Tech-
  nischen Hochschule
Aachen, Germany

Jose Katz
Institut für Theoretische
  Physik
Freie Universität Berlin
Berlin, Germany

Abraham Klein
Department of Physics
University of Pennsylvania

Hans Kleinpoppen
Department of Physics
University of Stirling
Stirling, Scotland

Behram Kursunoglu
Center for Theoretical
  Studies
University of Miami

G. L. Lamb, Jr.
United Aircraft Research
  Laboratory
East Hartford, Connecticut

Willis E. Lamb, Jr.
Department of Physics
Yale University

Peter Lambropoulos
J. I. L. A.
University of Colorado

Neal F. Lane
Department of Physics
Rice University

Francesco E. Lauria
Istituto di Fisica Teorica
Naples, Italy

Benjamin Lee
Institute of Theoretical
  Physics
State University of New
  York at Stony Brook

Bernard A. Lippmann
Department of Physics
New York University

Chun-Chian Lu
Center for Theoretical Studies
University of Miami

Joseph Macek
Department of Physics
University of Nebraska
  at Lincoln

Harrie S. W. Massey, F.R.S.
Department of Physics and
  Astronomy
University College London
London, England

V. S. Mathur
Department of Physics and
  Astronomy
University of Rochester

Samuel McDowell
Department of Physics
Yale University

Rui Vilela Mendes
Instituto de Fisica e
  Matematica
Lisbon, Portugal

LIST OF PARTICIPANTS

Sydney Meshkov
National Bureau of
  Standards
Washington, D. C.

Stephan L. Mintz
Center for Theoretical
  Studies
University of Miami

Robert S. Mulliken
Department of Physics
University of Chicago

Richard E. Norton
Department of Physics
University of California
  at Los Angeles

Robert J. Oakes
Department of Physics
Northwestern University

Robert F. O'Connell
Department of Physics and
  Astronomy
Louisiana State University
  at Baton Rouge

Reinhard F. A. Oehme
Enrico Fermi Institute
University of Chicago

Sadao Oneda
Department of Physics and
  Astronomy
University of Maryland

Lars Onsager
Center for Theoretical
  Studies
University of Miami

Konrad Osterwalder
Lyman Laboratory of
  Physics
Harvard University

Jogesh C. Pati
Department of Physics and
  Astronomy
University of Maryland

Arnold Perlmutter
Center for Theoretical Studies
University of Miami

Arnulf Rabl
Department of Physics
Ohio State University

Mario Rasetti
Center for Theoretical Studies
University of Miami

Tullio Regge
Institute for Advanced Study
Princeton, New Jersey

E. Reichert
Institut für Physik
Johannes-Gutenberg-Universität
Mainz, Germany

Rudolf Rodenberg
Physikalisches Institut der
  Rheinisch-Westfälischen
  Technischen Hochschule
Aachen, Germany

Robert Rosen
Department of Electrical
  Engineering and
  Systems Science
Michigan State University

S. Peter Rosen
Department of Physics
Purdue University

D. P. Roy
Rutherford High Energy
  Laboratory
Chilton, Didcot, England

Bunji Sakita
City College of the City
  University of New York

LIST OF PARTICIPANTS

Abdus Salam
International Centre for
  Theoretical Physics
Trieste, Italy

Y. R. Shen
Physics Department
University of California
  at Berkeley

Ralph L. Sherman
J. M. Richards Laboratory
Grosse Point Park, Michigan

James A. Slevin
Department of Physics
University of Stirling
Stirling, Scotland

Kenneth Smith
University of Leeds
Leeds, England

Larry Spruch
Department of Physics
New York University

Andrew Stewart
Center for Theoretical
  Studies
University of Miami

E. C. G. Sudarshan
Center for Particle Theory
University of Texas
  at Austin

Takehiko Takabayasi
Department of Physics
Nagoya University
Nagoya, Japan

Aaron Temkin
NASA Goddard Space
  Flight Center
Greenbelt, Maryland

Dietrick E. Thomsen
Science News
Washington, D. C.

Ronald Torgerson
Department of Physics
Ohio State University

Hiroomi Umezawa
Department of Physics
University of Wisconsin
  at Milwaukee

Kameshwar Wali
Department of Physics
Syracuse University

Steven Weinberg
Department of Physics
Massachusetts Institute
  of Technology

Dwight L. Wennersten
Air Force Office of
  Scientific Research
Arlington, Virginia

A. S. Wightman
Department of Physics
Princeton University

Robert R. Wilson
National Accelerator
  Laboratory
Batavia, Illinois

Jerome Wolken
Biophysical Research
  Laboratory
Carnegie-Mellon University

Armand Wyler
Mathematical Institute
University of Zürich
Zürich, Switzerland

Fredrik Zachariasen
Division of Physics
  and Astronomy
California Institute of
  Technology

SUBJECT INDEX

Action
    and Weyl's geometry   11
    gravitational part   11
    invariant in Einstein theory   11
Annihilation of positron
    angular correlation of two γ-decay   192
    cross-section   191-192,194,200,203,206-210
    experimental detection   194
    ore gap   194
    orthopositronium   207
Biology
    bifurcation set   376
    biochemical determinism   362-372
    catastrophe set   376
    evolution of biological system   357
    morphogenesis   358-361
    quantum structural stability   385
Charge conjugation   38
Chiral algebra   175
Co-scalar   9
Cosmology
    spin-isotopic-spin-tortional effects   82
Co-tensor   9
Current density   49
Curvature tensor   10
Detectors
    magnetic spectrometer   109-110
    NaI   120
    TANC   110,114
    TASC   110,114,120
    total absorption   109
Dirac and Kemmer Equations   42
Duffin-Kemmer Matrices (generalization of)   26
Elastic continuum
    including interaction   173
    relativistic mechanics of   169-173
    tachionic states   173
Electromagnetism
    formulation of   20
    waves   21
Electron-Atom Collisions
    direct differential cross section   233-237,247-250
    exchange differential cross-section 233-237,247-250
    spin polarization by Hg   250
    spin polarization by Ne   231
Electron-Molecule Collisions
    adiabatic-nuclei theory   298-305

Frommhold rotational resonances 285-298
impulse approximations 271-283
resonances in diatomic molecules 262-265, 306-312
shape resonances in H_2 265-266
shape resonance model 267-270
Energy-momentum tensor 25
f^o meson 56
Field Equations 62
Cartan's tortional 62
Einstein's curvature 62
tetrode identity 62
Gauge Theories
and non-compact $SL(2,C)$ 57
Einstein-Cartan-Weyl Theory 57
E-M mass difference 139
details of hadron model 131
introduction of non-strong interactions 133
M particles 130, 138
renormalizability of spontaneously
broken 127, 129
scaling 139
Geometry and physics 87
G-M-O Formula 175-177
Gravitational constant
and the age of the universe 7
modification of Einstein's theory 8
weakening with time 7
Groups
broken $SU(3)$ 175
G_o 31
G_I^o 39
$SL(2,C)$ 56
$SL(6,C)$ 55
$SO(3,2)$ 26, 28
$SU(3)$ 56
$SU(6)$ 85
$SU(3,1)$ 24, 28
Helium in small assembly
Bernoulli equation 340
density wave 343
dispersion 338, 341-342
energy of a droplet 348-351
energy of a slab 346-347
Huilier formula 340
non-local enthalpy 341, 343
nuclear fission model 339, 351-352
roton 339

SUBJECT INDEX

surface tension	344
Yukawa potential	347
Higgs mesons	126
In-scalar	9
In-tensor	9
Intermediate boson	53,164
Internal spin-torsion	57
Lagrangian	
density	53,183
Einstein-Cartan gravitational	56
gauge	58
invariance under SL(2,C)	56
invariance under SL(6,C)	56
Locality	93
Lorentz transformation	24
Mass of elementary particles	
formulae	184
restrictions imposed by gravity	89
spectrum	53,54
Maximon	90,92-93
NAL	
accelerator	98
neutrino area	103
production of gamma rays by pi^{o}	101
proton scattering experiments	99-102,110
site	98
Nonet	
of 2+ particles	56,76
of 2⁻ particles	76
Ordering of events	83
Parallel displacement of vectors	
electromagnetic potentials from	4
non-integrable lengths of vectors	4
quantum effects	5
Weyl's generalization	3
Photo-ionization	
Fano-effect	238-244
spin polarization from multiphoton processes	244,254-256
Photon	
role in quantum theory	19
polarization	20,21
Pions	120
pi-N interactions	145
pion mass difference	152-161
pi-pi interactions	145
recent evidence on charge independence	161-165
Point-like events	86
Positron Collision in Gases	
experimental apparatus	204

SUBJECT INDEX

momentum transfer cross-section	192
source energy distribution	198
total cross-section	192-195,201,202
transmission experiments	196
Quantum completion	72-79
Scalar-Tensor Theory	12
development of	13-14
Spin polarization	
from unpolarized target and projectiles	250-253
(see electron-atom collision)	
(see photo-ionization)	
SPEAR	120
Stochastic metric	85
Symmetry breaking	
and past and future	15
and sign of charge	15
P, C, and T	16
Unification	
of gravity and electromagnetism	6
of spin and internal symmetry	57
Variational principle for atomic scattering	
adiabatic approximation to variational bound	219-222
linear Hermitian operator	222-226
velocity dependence of exchange cross-section	215-219
Vierbeins	67
72-beins	67
W boson	53,104
mass formula for	54
Weak and strong interactions	41
Weak discharge Penning processes	246-247
Weyl Connection	60
X-ray from Galaxies	
Andromeda	327
binding and missing mass	331-332
black hole	318
Cen A	323-324
Cen X-3	322-323
clusters of galaxies	331
coma cluster	330
Crab Neubla	318-320
Cygnus X-1	322-323
death of stars	317
diffuse x-rays	326-329
Her X-1	322-323
inverse Compton scattering	327
luminosity of galaxies	317
M-87	323
matter and antimatter	332-335

SUBJECT INDEX

```
Perseus A                              323
pulsars                            317,328
quasars                            325-326
Sco X-1                            320-321
Seyferts                           325,328
supernova                              317
very soft x-ray                    329-331
virial theorem                         331
```